Democratizing No-Code Application Development with Bubble

A beginner's guide to rapidly building applications with powerful features of Bubble without code

Caio Calderari

Democratizing No-Code Application Development with Bubble

Group Product Manager: Aaron Tanna

Publishing Product Manager: Uzma Sheerin

Book Project Manager: Deeksha Thakkar

Senior Editor: Nisha Cleetus

Technical Editor: Rajdeep Chakraborty

Copy Editor: Safis Editing

Proofreader: Nisha Cleetus

Indexer: Manju Arasan

Production Designer: Vijay Kamble

Business Development Executive: Puneet Kaur

DevRel Marketing Coordinators: Deepak Kumar and Mayank Singh

First published: April 2024

Production reference: 1190424

Published by Packt Publishing Ltd.
Grosvenor House
11 St Paul's Square
Birmingham
B3 1RB, UK

ISBN 978-1-80461-094-7

www.packtpub.com

To my wife, Maisa, for being my loving partner throughout our joint life journey. For all the love, partnership, help, and support during the writing process and our entire life together. To my parents and sister for the inspiration, education, and support that allowed me to achieve my goals and grow as the person I am today. To my parents-in-law, who took me in as part of their family, providing support, help, love, and kindness. Without you all, nothing would make sense.

– Caio Calderari

To the creators, innovators, thinkers, builders, and doers. To the ones who want to make a difference, create something new, build their dreams and ideas, and just be able to do it. To the empowered people who will be able to change their realities and their surroundings and impact people's lives in a positive and transformative way. To the builders who will change and shape the world through the art of making their ideas a reality. To the ones who will make it happen.

– Caio Calderari

Contributors

About the author

Caio Calderari is a No-Code Expert, Startup Founder and Senior Product Designer with 17+ years of experience. Starting as a designer in 2005, he worked for advertising, digital agencies, corporate companies and startups worldwide.

He learned no-code tools to be able to build his own startups and side projects and later decided to teach others too. In 2020, he created a YouTube channel about no-code tools offering courses and tutorials. By 2021, he became the Chief No-Code Officer at a U.S.-based company, helping entrepreneurs from various countries launch their startup ideas.

Caio is a creative individual, passionate about sharing his knowledge. His goal is to bring no-code to more creators, helping them unlock this new world of possibilities.

If you want to connect with Caio Calderari, you can follow him on the social networks with @calderaricaio or visit his website titled Caio Calderari - No-Code, Design & Startups.

About the reviewer

Chris Schrade is a seasoned no-code developer and Founder at one of the USA's leading Bubble.io Agencies, Alchemy Apps. His journey began in 2016, aiming to streamline workflows for his employer via an internal application. Discovering Bubble transformed his approach, leading him to be part of 100+ applications. Beyond his agency work, Chris's passion for technology and innovation spurred the creation of side projects like OUTWRK, a workout log, social platform, and marketplace for athletes, coaches, and parents; Market Summary Assistant, an AI-powered stock market summarizer; and Ooldie, a service connecting seniors with in-home care providers and the tools to manage the relationship. His commitment to simplifying technology for practical, impactful use is evident through his work and contributions to the no-code community.

Table of Contents

3

Building Blocks – Exploring Bubble's UI Components   53

4

Building Your First Bubble App – The Planning Phase   89

8

Database Structuring, Relationships, and Security   167

9

Extending Functionality with Plugins and APIs   193

10

Testing and Debugging Strategies   221

11

Deploying and Launching Your App (Publishing)   241

12

Monitoring, Maintenance, and Updates (Apps Governance)   251

13

Optimizing Performance and Scalability   263

Preface

In today's world, every company is a software company and needs to embrace technology. High and growing demand for tech solutions meets a shortage of skilled developers worldwide posing a hiring expenses challenge for companies.

The digital landscape is expanding at an unprecedented pace, and those who fail to incorporate technology into their operations risk being left behind or dying. Digital transformation is a strategic competitive advantage that can't be ignored.

The key to thriving in this competitive tech-driven era lies in the rapid and cost-effective creation of software that delivers value to customers.

No-code tools offer a faster, more affordable software development alternative with a transformative approach to software development that allows more people to be able to build software. This revolution is allowing common people and company's employees, with no prior technical skills and programming knowledge, to be able to build software faster, easier and more affordable.

No-Code, Low-Code Tools and Artificial Intelligence are groundbreaking solutions, enabling individuals to craft applications without the need for years of coding expertise. This is a game changing scenario that can help with the existing challenges we face in the software world. With intuitive drag-and-drop components and plain English logic, these tools open doors for learners of all backgrounds to embark on their software-building journey in a matter of months, not years.

During the pandemic No-Code tools became even more relevant, helping businesses migrate their physical operations to the internet, becoming digital, in a fast and affordable way. Since then, these tools have gained more market traction, investments and evolved over the years.

Leading the charge in the no-code revolution is Bubble, which represents one of the pioneers in the No-Code Revolution and that has one of the largest community and user base with more than 3 million users worldwide. Bubble is a powerful platform that empowers newcomers to construct remarkable web applications in a matter of weeks if not days.

Its accessibility has empowered companies to devise internal solutions, entrepreneurs to forge the path for their startup MVPs, and creatives to offer bespoke development services to clients. This book is an introduction to the no-code revolution, it is a way for creative people to be able to express their creativity by being able to build their dream solutions themselves, to build software and accomplish tasks, solve problems. As I discovered this amazing world of possibilities, I hope I can help you along the way by providing a little bit of help and guidance in your learning path.

Everyday individuals, driven by their ideas, wanting to change their reality and the world, can now turn them into reality without the steep learning curve traditionally associated with coding. This is the No-Code Revolution, and I am glad you are now a part of it. Welcome. Let's No-Code!

Who this book is for

This book is for Aspiring No-Code Developers, Citizen Developers, Startup Founders, First Time Entrepreneurs, Designers, and Makers who want to learn, in a step-by-step manner and without using code, how to build web applications with Bubble, a no-code tool that can allow you to build a high-quality web application or website to serve your specific business objectives.

What knowledge do you need before getting started? You don't need technical skills, you don't need to know how to code or program, or how to design websites, apps or digital software. That is the beauty of no-code tools, it allows basically anyone to learn and be able to build digital applications. But if you do come from a technical background or have web design and development skills, that is also fine and helpful.

It is recommended you have at least familiarity with how the internet works, what a website is, how to install and use an app and how to use basic software like Spreadsheets and PowerPoint, for instance. This book is targeting beginners, so it won't require any initial and specific technical knowledge or skill. If you know how to operate a computer and use the internet and common software, you will be ready to read the book and learn.

What this book covers

Chapter 1, Getting Started with Bubble.io - Exploring Bubble's Features, This chapter is an introduction to Bubble, a powerful no-code platform for building web applications in a visual way. Readers will discover the fundamentals of the tool, examples of what you can build with it. Learn and explore its essential main features, like the visual editor. Practically learn step-by-step how to set up an account and workspace, and gain familiarity with the main and most important areas of the tool allowing them to get started with Bubble and no-code to build their first projects in a visual way, without code.

Chapter 2, Navigating the UI Builder Components Tab, This chapter will explore the UI Builder Components Tab, a crucial aspect of building web applications on Bubble.io. Readers will learn fundamental concepts such as the Elements Tree and its role in managing component visibility and conditional settings. Through practical examples, they'll grasp key features of container components, input forms, and reusable elements, essential for crafting intuitive user interfaces. Additionally, readers will learn how to expand their toolkit by installing additional components from the Plugins Marketplace and gain valuable insights into efficient component management practices. By understanding these core features, readers will be equipped to navigate the UI sidebar with confidence and seamlessly build their own applications on Bubble.io.

Chapter 3, Building Blocks - Exploring Bubble's UI Components, This chapter will cover the UI components available in Bubble, giving a comprehensive understanding of each element's purpose and functionality. From essential visual elements like Text, Button, and Image to versatile containers such as Group and Popup, readers will learn how to utilize these building blocks to craft compelling page layouts for their digital applications. Additionally, through clear explanations and practical examples, readers will discover the power of reusable elements and learn how to efficiently incorporate them into their projects. By the end of this chapter, readers will be equipped with the knowledge and skills to confidently navigate Bubble's UI components and leverage them effectively in their own application development journey.

Chapter 4, Building Your First Bubble App – The Planning Phase, In this chapter, readers transition from exploring Bubble's components to the essential planning phase of app development. Readers will learn skills to strategically plan their no-code applications, ensuring clarity and efficiency throughout the development process. By identifying target users, defining project goals, and outlining desired functionalities, readers lay a solid foundation for their projects, minimizing complexity and streamlining development. By the end of the chapter readers will have gained a holistic understanding of the planning process, setting the stage for successful app development.

Chapter 5, Layout and Styles, This chapter covers layout and style customization options. Readers will learn how to work on the visual aesthetics of their applications. By navigating the styles tab, readers gain essential skills to customize elements according to their preferences, including colors, typography, and layout structures. Understanding these customization options empowers readers to design beautifully crafted applications aligned with their brand's style and guidelines. Through layout and design adjustments, readers enhance the overall user experience, creating visually appealing interfaces that captivate and engage users.

Chapter 6, Building User Interfaces with Bubble, In this chapter, readers will learn about responsive layouts and user interface design. Building upon the foundation established in the previous chapter on layouts and styles, readers will learn how to create responsive layouts that seamlessly adapt to diverse devices and screen sizes. Through exploration of Bubble's responsive design features within the responsive editor, readers acquire the skills to craft visually appealing and user-friendly interfaces. By mastering responsive layout creation and configuration, readers ensure their applications are accessible and optimized across various devices.

Chapter 7, Workflow Automation and Logic, This chapter introduces workflows and logic features, empowering readers with the skills to implement logic in their Bubble applications. Readers will learn how to configure workflows in elements, creating actions and triggers. Workflows serve as the essential "brain" of any Bubble application, orchestrating interactions between the front end and the database. By integrating front-end designs with the database through workflows, readers will be able to develop powerful applications efficiently. Additionally, readers will learn about back-end workflows, conditionals and logic statements, unlocking the potential for building advanced automations within their projects.

Chapter 8, Database Structuring, Relationships and Security, This chapter focuses on database structuring, relationships, and security within Bubble applications. Readers will learn what a database is, how to create one and establish relationships between data elements. Additionally, they will learn how to implement security measures to safeguard their database information. Through practical examples, readers will learn how to create databases, define data types and fields, and utilize workflows to populate databases with user information. By mastering the integration of databases and workflows, readers will acquire the skills needed to develop dynamic applications in Bubble.

Chapter 9, Extending Functionality with Plugins and APIs, This chapter is about Plugins and APIs. Readers will learn how to explore Bubble's plugin ecosystem, the marketplace, learning how to seamlessly integrate their applications with external services using APIs and Plugins. They will also learn how to find, manage and install new plugins. Additionally, they will also learn about APIs and the API Connector plugin that allows integration between Bubble and any existing API from a third-party service. By the end of this chapter, readers will be empowered to expand the capabilities of their web applications, with the power of APIs and Plugins to accomplish a diverse range of tasks such as adding new components, collecting payment, using an external database and much more.

Chapter 10, Testing and Debugging Strategies, In this chapter readers will learn about existing features inside Bubble that allow them to test and debug their applications. They will also learn other methods and ways to test and debug their application, including various types of tools and user testing. This is a practical guide to testing and debugging your application effectively. By mastering these techniques, they will be equipped to identify and resolve issues, ensuring a seamless user experience and application performance.

Chapter 11, Deploying and Launching Your App (Publishing), In this chapter, readers will learn how to deploy and launch their application in a practical way. They will learn about the deployment process, from thorough preparation to the actual launch, with a practical checklist ensuring a seamless and successful deployment experience. By the end of this chapter, they will be equipped with the necessary skills to confidently navigate the deployment process, including revising, previewing, and testing their app. Additionally, they will learn how to set up a custom domain and effectively launch your app to the public, making it accessible to users worldwide.

Chapter 12, Monitoring, Maintenance, and Updates (Apps Governance), This chapter will teach about app management, focusing on monitoring, maintenance, and updates to ensure the optimal performance of an existing application. It will cover Workload Units (WUs) and explain how these metrics quantify resource utilization and impact operational costs. Readers will also learn about version control, updates, and app governance, vital for effective management of no-code applications. Monitoring app performance and user analytics, alongside addressing maintenance and user feedback, will be emphasized to ensure sustained success post-deployment. By the end of this chapter, readers will be equipped with the knowledge and tools to effectively manage and optimize your app for long-term success.

Chapter 13, Optimizing Performance and Scalability, The reader will learn about performance and optimization and how to improve the application as the user base starts to grow. The chapter covers tips on how to improve the app's overall performance, scalability, and the user experience. It will cover some of the best practices and provide advice for efficient app design, performance and scalability covering strategies to handle databases, workflows and creating interfaces that resonate with users and provide a pleasant experience which also impacts on the perceived performance of your application.

In this book we covered essential topics for you to learn Bubble and get started in your no-code journey building digital applications in a visual way. We also covered about project planning, user experience, performance and optimization. Finally, life cycle management concerns particular to Premium capacity features were explored.

This book has been all about how to enable you to enter the no-code universe and start building your own app ideas on your own. We hope you have enjoyed this book and are now confident in how to apply this knowledge to build amazing digital applications with no-code. Welcome to the No-Code space, welcome to the no-code revolution!

In this book, you've entered the world of Bubble and no-code development, learning the essential knowledge and skills to embark on your no-code journey with confidence. We've started with the basics of the Bubble, guiding you through the possibilities and examples until the creation of your first account and project. You also learned about interfaces, layout and style techniques for crafting responsive and visually appealing applications, this book has equipped you with the skills needed to understand the foundations about design and how to manipulate layouts effectively. You've learned how to enhance front-end functionality by implementing conditionals and logic to manipulate elements dynamically, ensuring an engaging user experience. Additionally, you've gained insights into managing databases securely, controlling access to public and private data to safeguard sensitive information.

We've explored backend functionality and workflows to automate processes and streamline operations. Furthermore, you've explored the integration of APIs, learning how to seamlessly connect Bubble with third-party services to extend the capabilities of your applications. You've uncovered basic tips and strategies to enhance the performance of your apps, ensuring smooth and efficient operation even under heavy usage. From minimizing load times to optimizing database queries, you now have the tools to create high-performing applications that deliver a seamless user experience.

In addition to these core topics, you've also gained insights into project planning, user experience design, and project management. By covering a wide range of essential topics, we've provided you with a comprehensive foundation to kickstart your journey in building digital applications visually using Bubble and No-Code.

This book is about much more than just learning Bubble. It's about empowering you to unleash your creativity, turn your ideas into reality, and make a meaningful impact in the digital landscape. It is about the democratization that comes with no-code and tools like Bubble, allowing non-technical and technical people to work faster and more effectively building software that can help other people and solve real problems. We've provided practical guidance, step-by-step instructions, and valuable insights to help you navigate the complexities of app development with ease.

To get the most out of this book

Create a Bubble account. The free plan will work for beginners. If you wish to publish your app and test premium features, an upgrade is recommended.

Read the book and put learnings into practice, your actual learning will come from using the tool, so make sure you apply the knowledge in a practical way.

Research more about topics and subjects you are interested in. Bubble is a big and complex tool, it would be very hard to cover everything needed in a single book, so it is recommended you keep learning as you go and search for more content during your no-code learning journey.

Bubble is a web tool that runs on your browser. You won't need to install anything on your computer or update it. To use it, simply go to their website `https://bubble.io` *and create an account, log in and start building. Updates will happen automatically, you won't need to do anything, unless the Bubble team does tell you to do anything to update your app. You also won't need to save your app and work in progress when using the editor, changes are automatically saved, just make sure you are online.*

To use Bubble all you need is an account, a simple computer and internet connection, that's it.

Software/hardware covered in the book	Operating system requirements
Bubble	Windows, macOS, or Linux
Google Chrome	Windows, macOS, or Linux

Disclaimer:
Please note that all screenshots from this book are taken at the time of writing. They may vary due to updates made to the Bubble interface.

Conventions used

Bold: Indicates a new term, an important word, or words that you see onscreen. For instance, words in menus or dialog boxes appear in bold. Here is an example: "You can also check the **My Templates** item under your account."

Tips or important notes
Appear like this.

Get in touch

Feedback from our readers is always welcome.

General feedback: If you have questions about any aspect of this book, email us at `customercare@packtpub.com` and mention the book title in the subject of your message.

Errata: Although we have taken every care to ensure the accuracy of our content, mistakes do happen. If you have found a mistake in this book, we would be grateful if you would report this to us. Please visit `www.packtpub.com/support/errata` and fill in the form.

Piracy: If you come across any illegal copies of our works in any form on the internet, we would be grateful if you would provide us with the location address or website name. Please contact us at `copyright@packt.com` with a link to the material.

If you are interested in becoming an author: If there is a topic that you have expertise in and you are interested in either writing or contributing to a book, please visit `authors.packtpub.com`.

Share Your Thoughts

Once you've read *Democratizing No-Code Application Development with Bubble*, we'd love to hear your thoughts! Scan the QR code below to go straight to the Amazon review page for this book and share your feedback.

`https://packt.link/r/1804610941`

Your review is important to us and the tech community and will help us make sure we're delivering excellent quality content.

Download a free PDF copy of this book

Thanks for purchasing this book!

Do you like to read on the go but are unable to carry your print books everywhere?

Is your e-book purchase not compatible with the device of your choice?

Don't worry!, Now with every Packt book, you get a DRM-free PDF version of that book at no cost.

Read anywhere, any place, on any device. Search, copy, and paste code from your favorite technical books directly into your application.

The perks don't stop there, you can get exclusive access to discounts, newsletters, and great free content in your inbox daily

Follow these simple steps to get the benefits:

1. Scan the QR code or visit the following link:

https://packt.link/free-ebook/9781804610947

2. Submit your proof of purchase.
3. That's it! We'll send your free PDF and other benefits to your email directly.

Getting Started with Bubble.io – Exploring Bubble's Features

Welcome to the No-Code Revolution! This is an exciting change in the software development industry that will empower and allow more individuals to be able to build software, without having to learn how to code and without becoming traditional developers. Bubble is one of the most powerful no-code platforms available on the market, and one of the most popular as well, with millions of users already using it.

In this book, you are going to learn the fundamentals of this amazing tool, and with the power of no-code, I am sure you will also be able to build web apps visually, without learning how to code.

In this chapter, you will start by learning what Bubble is and how to set up your first account. You are also going to learn Bubble's main features and how to navigate inside the visual editor. With this essential knowledge, you will be ready to dive into the following chapters and learn how to apply the powerful features available inside Bubble.

This chapter consists of the following knowledge:

- What Bubble is and important information about the tool
- What you can build with Bubble – examples
- How to practically set up an account and your project workspace
- Understand the Bubble editor and how to navigate the interface

Introduction to the Bubble.io platform

Bubble is a powerful, robust, and flexible platform that enables anyone to build web applications visually, meaning you can learn how to use it and build software without learning how to code or spending years trying to become a developer.

The no-code revolution is empowering people around the globe and democratizing software development by giving its users the ability to quickly learn and build digital applications in a simplified way.

Start-up founders, entrepreneurs, employees, individuals, and organizations can utilize the power of no-code tools such as Bubble to save time and money while developing powerful web applications visually, without having to learn code, and that is a meaningful change in the software industry.

Founded in 2012, Bubble is an all-in-one, multi-purpose tool that acts as a visual programming tool and a cloud platform. This means you can use it not only to build your application visually, but can also put it to work hosting and deploying your application to the web, so you don't have to deal with hosting and infrastructure to run your application somewhere else. You can use Bubble for creating applications, while the cloud platform hosts and operates them. You design, construct, manage, and expand your app through the application editor, accessible directly in your browser without any downloads. Bubble runs on the web, and you use it in your browser window. It automatically saves your work as you build – all you have to do to use it is to be online. Once an app is built on Bubble, it resides in Bubble's cloud infrastructure, accessible from anywhere.

Bubble empowers you to craft web applications accessible via browsers on computers, tablets, and phones. These applications feature a database, enabling users to register, store, retrieve, and modify data. Bubble's versatile editor is open-ended, akin to traditional programming languages—there's no predefined limit to what you can achieve. By combining diverse data and logic operations, you can create applications ranging from basic to intricate. The key capabilities of Bubble are as follows:

- Establishing user accounts
- Managing data (saving, editing, deleting, and retrieving it)
- Real-time updates
- Integration with various external services through APIs (such as payments, data sources, and authentication providers)
- Crafting responsive apps that adapt to screen sizes

Here are some examples of what you can build with Bubble:

- **Web apps**: Create interactive and dynamic web applications for a wide range of purposes
- **Social networks**: Build your own social networking platform with features such as user profiles, posts, comments, and connections
- **Marketplaces**: Develop online marketplaces where users can buy and sell products or services
- **CRM systems**: Design custom **customer relationship management** (**CRM**) systems to manage contacts, sales, and interactions
- **Project management tools**: Craft tools for organizing tasks, projects, and teams, with features including task lists, timelines, and collaboration
- **E-commerce platforms**: Build online stores with product listings, shopping carts, payment gateways, and order management

- **Booking and scheduling apps**: Create appointment booking systems, event calendars, and scheduling tools

- **Educational platforms**: Develop e-learning platforms with courses, quizzes, progress tracking, and user engagement features

- **Content sharing platforms**: Build platforms for sharing articles, videos, images, and other multimedia content

- **Job boards**: Create platforms for posting job listings, connecting job seekers with employers, and managing applications

- **Real estate websites**: Develop property listing websites with search filters, property details, and contact forms

- **Healthcare applications**: Build apps for managing patient records, scheduling appointments, and medical information sharing

- **Financial tools**: Design budgeting apps, expense trackers, investment calculators, and financial management tools

- **Community forums**: Develop online discussion forums with threads, replies, user profiles, and moderation features

- **Membership sites**: Create platforms for offering exclusive content, courses, and resources to members

- **Crowdfunding platforms**: Build websites for crowdfunding campaigns, donations, and fundraising projects

- **Travel and booking websites**: Develop platforms for booking flights, hotels, tours, and travel packages

- **Event management apps**: Create tools for planning, promoting, and managing events, conferences, and seminars

- **Analytics dashboards**: Design data visualization dashboards to display and analyze metrics and insights

- **Custom business tools**: Develop tailor-made solutions for specific business needs, such as inventory management, employee tracking, and more

For the healthcare sector and other types of applications that require a HIPPA-compliant database or might store user's sensitive information, it is important to note that Bubble's internal database isn't recommended. If you plan to build such apps, it is advised to use an external database service that follows the guidelines and data protection laws of the countries that your app will serve.

There are several compelling reasons to invest your time and effort in learning Bubble. Firstly, it eliminates the need for coding expertise, allowing you to create sophisticated web applications through a visual interface. This accelerates your learning curve and democratizes app development, making it accessible to a broader audience.

Secondly, Bubble's rapid prototyping capabilities enable you to swiftly test and refine your app concepts, saving valuable time during the development process. Moreover, Bubble's cost-efficiency is noteworthy, particularly for startups and small businesses with limited budgets, as it reduces the need for expensive development resources. Beyond financial benefits, mastering Bubble empowers you with autonomy over your app's creation, modification, and maintenance. This independence fosters innovation and empowers you to explore creative and entrepreneurial pursuits. The versatility of Bubble stands out, allowing you to craft a diverse range of applications tailored to your goals.

With a supportive community and regular updates, Bubble ensures you're equipped with the latest tools and practices in app development. Whether you're aiming to prototype MVPs, launch startups, or explore new career avenues, learning Bubble equips you with a unique skill set that holds significant potential in today's digital landscape.

Setting up an account and project workspace

To start leveraging the power of no-code, all you have to do is create your first account for free on Bubble's website:

1. Your first step is to access the website (`https://www.bubble.io`) and click the **Get started** button.

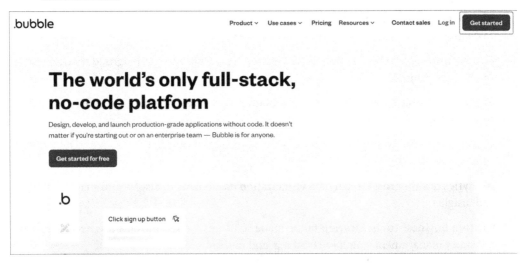

Figure 1.1: Bubble.io home page – the Get started button

2. To create your account, just fill out the form by adding your email and password. If you prefer, you can also sign up using your Google account (Gmail) if you have one.

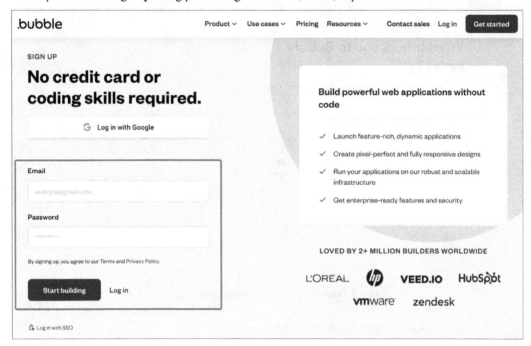

Figure 1.2: Bubble.io sign-up page

Follow the onboarding steps to complete your account profile information. These steps include a little survey to help Bubble understand more about you and your goals. The answers you choose here won't impact anything in your account.Just keep clicking until the entire onboarding is done.

> **Note:**
> These steps might change overtime, and it is expected, if the interface you see is different from the book, just continue the onboarding process until you get to the end and access your account. This part won't be difficult.

Figure 1.3: Bubble.io sign-up onboarding – step 1

1. Simply keep answering the questions to continue.

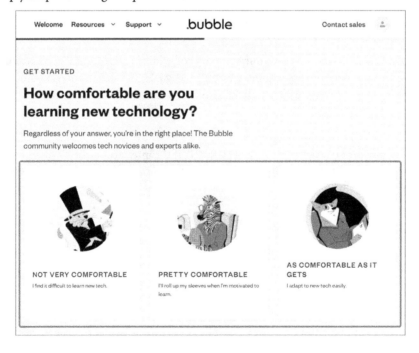

Figure 1.4: Bubble.io sign-up onboarding – step 2

2.   Continue clicking the options that fit your profile the best.

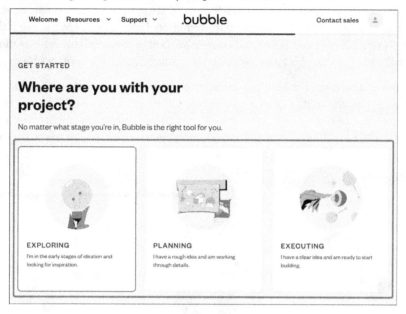

Figure 1.5: Bubble.io sign-up onboarding – step 3

3.   Choose an option shown in the following screenshot and continue.

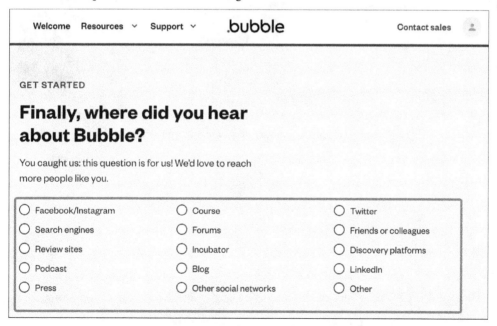

Figure 1.6: Bubble.io sign-up onboarding – step 4

4. After completing the onboarding process, it is important to check your email inbox, as you should have received a confirmation email. If you can't see it, you can wait a couple more minutes (sometimes there is a delay) or check your promotional and spam folders to see if the email is there. If you still can't find it, you can click on the button to resend this confirmation email. If this still doesn't work, check that you entered the correct email or contact support for help.

    You must open this email and click to confirm your account because this will unlock the next steps so you can start using Bubble. Once you click the email link, you will be sent to a Thank You page. You can close it and go back to the initial tab you were before you entered your email. You will notice the confirmation message will go away, unlocking you to continue with the next step.

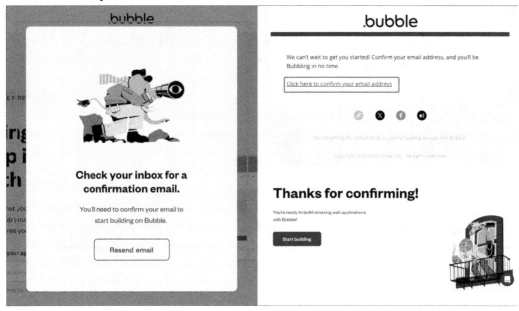

Figure 1.7: Bubble.io onboarding – email confirmation

5.   Once your email is confirmed Bubble will allow you to create your first project. To create a new project you can simply click the **Start building** button. But there is also an option that helps you build your project for the first time, with help of AI. You can type in the field a simple description of your app idea, which is optional, if you do fill out the field, it will generate your app and send you to a page with a few guidelines and tutorials on how to create your desired project.

Note that this feature might change overtime. For now I would recommend starting with an empty project, you can explore these AI features later while creating new projects. For the first one just click the Start building button and move forward.

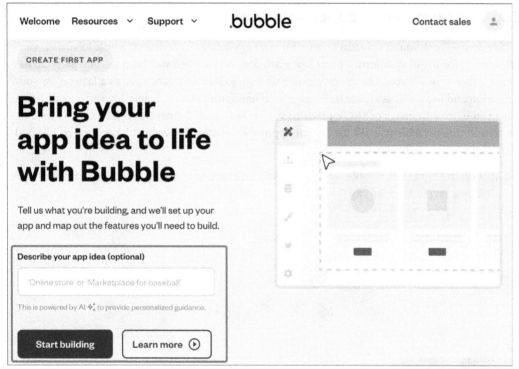

Figure 1.8: Bubble.io – Create the first project

Another way to start a project is by choosing a template, but we will cover more on that later.

Congratulations! At this point, you have successfully created your account. You also completed the onboarding process and started creating your very first application from scratch. Now it is time to dive deeper into how Bubble works and start configuring your app visually. In the next section you are going to learn about the application setup helper called application assistant. Let's go!

New project setup – application assistant

Every time you create a new project inside Bubble, you will be asked to set up this project for the very first time. This is the application assistant. Bubble will show the *get started* steps and ask you a few questions to make sure you have your initial settings ready to go. This process is helpful but can also be skipped. All the little details and settings you will be asked to define during these steps can also be found inside Bubble later, so if you skip them, you can always go back and change them later. Let's take a look at how the *getting started* steps appear.

Step 1 – favicon and application name

The first step will ask you to upload a favicon for your application. A favicon is a little icon, an image, that shows at the top of your browser window. Each page on your browser has a tab and each tab has an icon – this is the favicon. The image you upload here to Bubble will be used as a favicon for your application and users will associate this image with navigating your web app. My recommendation is that you upload an image of 16x16 pixels in size and in `.png` format to be used as your favicon. The second thing Bubble will ask for is your application name. If you don't have one yet, just add something and change it later.

Note that on the right side, Bubble is going to show an example of how your favicon and your app name will look on the browser tab.

Figure 1.9: Bubble.io – application assistant – step 1

Now, of course, if you don't want to continue answering these questions and want to just jump straight into the project, you can click the **Skip application assistant** button. Feel free to do so.

Step 2 – default font

Now Bubble will ask you to pick your project's default font. This is the primary font of your application, meaning it will be visible in texts, buttons, and so on. You can also use more fonts in the future and change the primary font as well. At this point, you can choose one option from the list provided by clicking the dropdown or just pick the default and continue to the next step.

Figure 1.10: Bubble.io – application assistant – step 2

Once you pick an option Bubble will show an example layout on the right side with the selected font so you can have an idea of how it will look.

Step 3 – default colors

During this step, you can define a few colors that will be used on your application including the **Primary color**, **Primary contrast**, **Text color**, **Surface**, **Background**, **Destructive**, **Success**, and **Alert**. Again, you can change the colors and see how they appear on the right side. Use the color picker by clicking an existing color or specify your own hex codes if you already have your own specific brand colors. If you don't know much about design, just use the default colors and change these later.

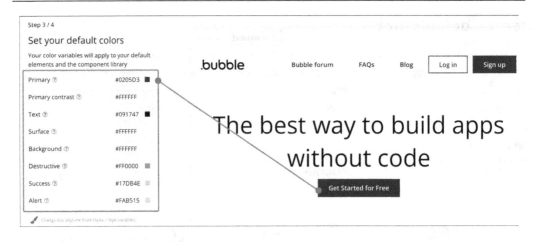

Figure 1.11: Bubble.io – application assistant – step 3

Once you are done configuring your colors, you can preview how it looks on the right side. If you are happy with your colors just click the **Next step** button to continue.

Step 4 – choosing plugins

Bubble has a lot of plugins you can install in your application to have extra features available for building amazing projects. These are like superpowers that let you build beyond what Bubble is capable of by itself. This step lets you pick plugins from a list visible on the right side, so you can save time by adding some plugins at the very beginning. If you wish to install a plugin, just click the **Install now** button. If you don't see a plugin you need on this list, you can click the **Browse all plugins** button, find the one you need, and install it that way.

If you know a specific plugin and know you are going to need it, this can be the perfect moment to install it. If you don't have any idea of what plugins you will use, I recommend you just skip this step and install plugins later as required. Remember, you can always go back to the plugins page to install or remove plugins.

Figure 1.12: Bubble.io – application assistant – step 4

Once you are done with this step, click the **Get started building** button to finish the application assistant process.

Finishing the application assistant

At this point, you have created your very first project and configured the initial settings using the application assistant. Next, Bubble will take you to your project and show you the Bubble editor.

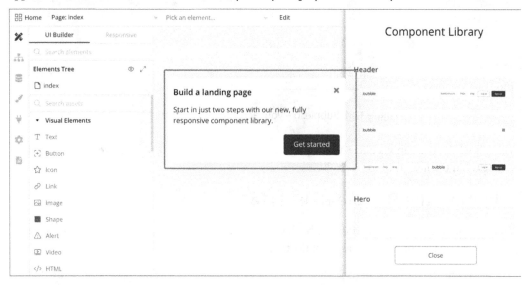

Figure 1.13: Bubble.io – the Bubble editor with a guided tour

Note that Bubble will continue to help you understand how the tool works by giving you a guided tour showing you around, what is what, and where things are. Following the tour is optional by clicking the blue buttons or you can choose to skip it. Your project has now been created and is ready to go, so you can start building! In the next sections, you will learn the essentials of the Bubble editor so you know where things are and how to navigate the tool.

With your account and project created, you now have the foundation to start building with Bubble. If you exit the project you just created, you will see your account workspace (dashboard) with all your projects side by side. To do that, just click the Bubble logo on the left-top corner of the page.

Here is an example of what your account and projects list looks like:

Figure 1.14: Bubble.io – your account and projects

This area is where you can visualize all the projects created under your account. We will cover it in greater detail in the next sections.

Navigating the Bubble.io interface

When using Bubble, your application will be developed inside the Bubble editor, so it is important that you get familiar with the main areas of the tool.

The editor is composed of a set of tools that allow you to access parts of your application, show/hide features, configure preferences, and define how your application should work. Let's discover the main areas of the Bubble editor.

The following screenshot shows the Bubble editor, where you build your applications visually.

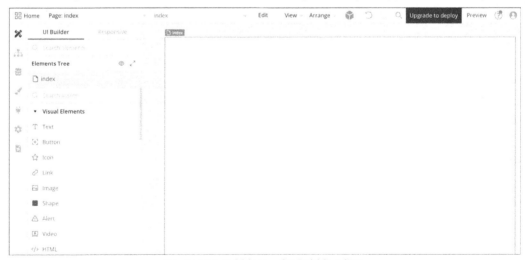

Figure 1.15: Bubble.io – the Bubble editor

The editor is composed of various areas. This big main area at the center is the canvas, where you will build your page layouts. You are going to use it a lot, but it is also important to know the other little parts of the editor and how to navigate to different areas using menus and tabs, so let's dive right in.

1 – the top bar

The top bar is a very important part of your editor and is always visible. From here you can go back to your account's projects, access the current project pages, configure visibility options to customize Bubble to your preferences, check that your project was saved, if there are any errors, and much more. One of the key features in the top bar is the preview button that allows you to visualize how your application looks in a new window.

Figure 1.16: Bubble.io – Bubble editor – top bar

2 – the sidebar

The sidebar is where you can find navigation tabs to access areas beyond the editor, such as workflows, data, styles, plugins, and more. At the top of the sidebar, you can also access responsive features, toggle your page to responsive mode, and visualize your page layers and components tree. It also shows all the components available for you to build your application, another very important part of the Bubble editor. The components sidebar is organized by categories. In each category, you can find specific components to be used to build your pages. All the components are *drag and drop* so you can easily select them and add them to the page. You will use the sidebar throughout your entire project build.

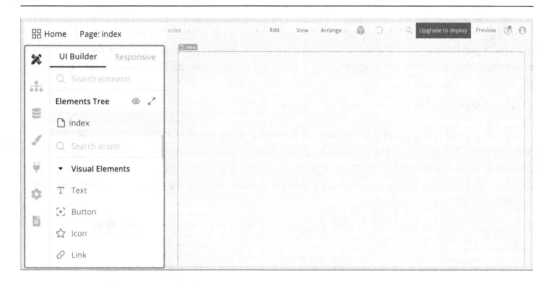

Figure 1.17: Bubble.io – Bubble editor – Sidebar

3 – the page

Inside the page is where you build your application layout by adding components and configuring their actions. If you were a painter, this would be your canvas. The page is a blank space where you will build your app's interface, the frontend part of your application. The cool part of using no-code is that you build things visually; you don't need to type code and then render the code to view the result. What you add on the screen and see as your page layout and design on that page is exactly what your users will see too. This is a much more creative, visual, and interactive way to build apps. As you may already have noticed, the page (canvas) is the heart of your application inside your Bubble editor.

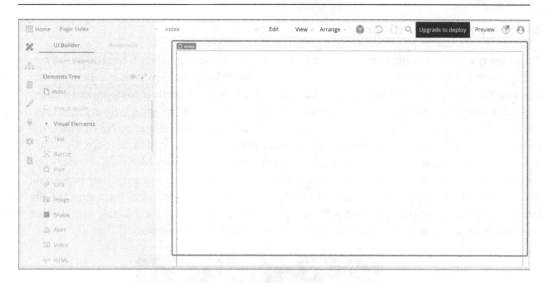

Figure 1.18 Bubble.io – Bubble editor – Page

Besides *page*, we can also call it the canvas, visual editor, or visual builder. When you are building your project inside the Bubble editor, you will see some things that your users won't see, such as guidelines and extra information that help you understand your layout structure. That is normal because you are visualizing your page as the builder, not the user at this moment. To view it as a user, click the preview button in the top bar to open your project in a new window. When you go to preview mode, the Bubble interface will no longer be there, so you will be able to navigate on that specific page as if a user was actually opening your application from their browser. By using the preview mode you get a more authentic navigation experience of that page, without the view of the editor getting in the way.

It is also common to have components not being clickable or not loading dynamic content inside the editor. Likewise, when you preview the page, it will load the components and make static components work, buttons will be clickable, and dynamic features and data from the database will function normally. So keep this in mind: the editor and page area are used to build, not navigate or test the application. The preview option is used to navigate and test the application.

4 – the property editor

The property editor is a floating panel that is used to configure your components. It can also be fixed on the right side of your editor if you prefer. You can do this by clicking the **View** item at the top bar and checking the **Lock property editor** option. The property editor is a contextual component, meaning you will see information about the component you are selecting. For instance, if no component is selected, the page is treated as the primary target and is automatically selected, so in this case, the property editor will show the features and settings you can edit that are related to that page. If you add a button to the page and you are currently selecting that button, the property editor will automatically adjust and show the features and settings related to that specific button, and this will keep happening during your application-building process. Basically, the property editor exposes the settings of any component you want to configure and will show you all the options you have available to make that component work as you wish. Let's see this in action in the following screenshot:

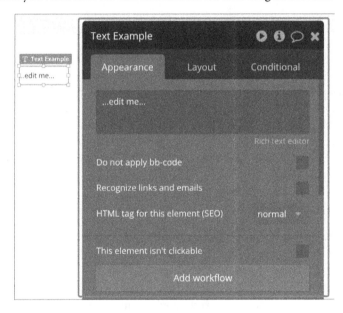

Figure 1.19: Bubble.io – Bubble editor – property editor

The property editor has a top bar where you will find the component name. Clicking this allows you to rename the component. On the right side, there are a few icons. The video icon displays a quick video explaining a bit more about the component you are using. The **i** icon shows more information about this element, and you can also access custom state settings there. The speech bubble icon is the notes area, which allows you to add custom notes to your component. This can be useful when collaborating with other people on Bubble or to just help you remember things in the future.

The second part of the property editor is the tabs. There are three tabs available that display different types of settings: **Appearance**, **Layout**, and **Conditional**. When you select an element, the property editor will appear and you can navigate through these tabs to access different areas and settings available for you to edit.

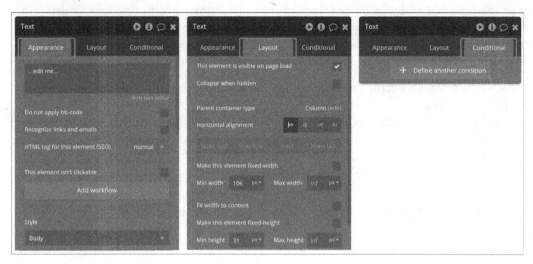

Figure 1.20: Bubble.io – Bubble editor – property editor tabs

Under the **Appearance** tab, you will find settings related to how the component looks and can change its colors, fonts, and styles. Under the **Layout** tab, you will find settings related to how the component behaves inside your page, its maximum and minimum size, whether it will stretch, be aligned to the left or right, its margins and paddings, and so on. Under the **Conditional** tab, you can create visibility rules to show or hide your component based on a variable condition or even change the style of that component if a user clicks a button, for instance. The **Conditional** tab allows you to work with dynamic data, as well as to create cool interactions, and even use custom states.

The property editor will follow you wherever you go inside Bubble and pretty soon it will be your best friend. Once you get the gist of it, you will understand how it works and how the fields adapt as you select new components and areas of the editor.

Note that some fields inside the property editor have fixed options for you to choose from, such as fonts, colors, and all sorts of fixed parameters, but other fields allow you to add dynamic data. This means you can add and create little formulas to display data coming from your database. We call this data binding. Basically, you write in plain English and in a logical way what type of information from the database you want to display inside a specific field, and with that you can bring dynamic data into components and build exciting things. But don't worry, let's not get ahead of ourselves. It might sound frightening and a lot at this point, but it will all make sense as we continue the book. Soon, things will start to connect and you will be ready to use Bubble in a simple and fluid way.

By now, you've learned about the main areas of your Bubble editor. This means you can now navigate around and understand how the tool is divided on screen. With this knowledge, it will be much easier to find things and play with the tool while building. It will also make you more familiar with Bubble, making it less intimidating to start. In the next section, you will learn about other areas of Bubble that are not initially shown to you and where you can find more interesting settings and features to play with to build amazing and powerful web applications.

Understanding the main areas of the editor and key features

In the previous section, you learned about the main areas available inside the Bubble editor. In this section, you are going to learn more about the main tabs and other internal areas inside Bubble in more depth so you understand what they are and how you can use them to configure extra settings and manage your own applications. Knowing each individual tab and what you can do inside them is an essential step because this will help you navigate inside Bubble and know exactly where you should go during your application building and development process. Let's dive right in.

Design

The **Design** area is the first tab on your Bubble editor. It is automatically pre-selected by default when entering a Bubble project. As the name suggests, this area is where you design your application – it is your visual canvas. Inside the **Design** tab, you visually construct your app's user interface by dragging and dropping components, resizing, positioning, arranging, and customizing elements such as buttons, text, images, inputs, and more to build your app's pages and layouts. If you are new to the platform, you are probably starting your journey here, playing with the visual components. Take some time to get familiar with this area as you will interact with it during your entire building process.

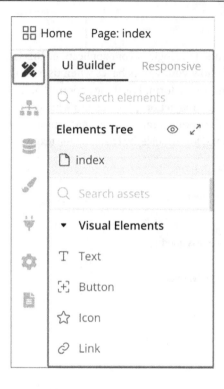

Figure 1.21: Bubble.io – Design tab

To get started, add a new fresh element to the canvas, so you can learn how it works. It is very simple. Click on a specific component in the sidebar located on the left, such as a button or text. Once you click the desired component, it indicates it is selected by changing the name of the component to another color. Now you can drag the chosen element onto the canvas (page). Or, you can move your mouse to the canvas and either click to add the item or even click and drag to draw the size of the component you want to add to the canvas. Thus, you have three options:

- Click and drag
- Click and click
- Click and drag while clicking

Now, to interact with the newly added component, you can double-click on the element to reveal the property editor. This will allow you to change this element's appearance and features. If you right-click on an element it will unveil a contextual menu offering supplementary editing possibilities.

So, to do a quick recap, in the **Design** tab, you can add components, define the layout of your app, change the settings of components and styles, and play with the responsiveness of these components to craft a polished and user-friendly design, all without the need for complex coding.

Workflow

The **Workflow** tab is where you build the logic of your application. This is the *brain* of your application. In this section, you can specify how your app behaves when users do things such as clicking on a button, viewing a page, or submitting a form. For these actions that users take on the frontend (the pages), you can create a set of rules and actions that react to what users did and perform changes on the pages they are visiting. These responses can change data, communicate with external services via APIs, or change how the page looks, for instance, by changing the text on a page or the color of a button.

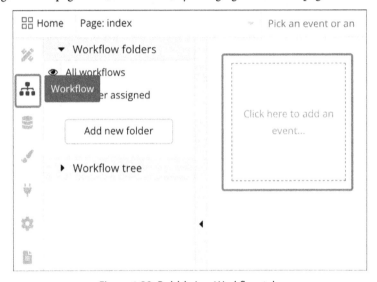

Figure 1.22: Bubble.io – Workflow tab

Workflows can seem a little complex at first, but this is one of the coolest parts when building with no code. Once you get familiar with it, it will be very exciting to play with workflows and create amazing functionalities for your web app. Don't worry, we will talk more about workflows in future chapters.

Data

The **Data** tab is where your database is. Bubble is an all-in-one no-code platform, meaning you can create a database and store information inside it. Of course, you can also choose to integrate Bubble with an external database, but if you are using the native database, then the **Data** tab is where you will find it.

Figure 1.23: Bubble.io – Data tab

This section lets you establish and manage the foundation of your app by defining data structures using data types and fields. It's where you create the tables and columns of your database, and fill them with real-world data. Inside the **Data** tab, you can set your database structure, create relationships, define and view all the items available inside your tables, and change and add fields and types. You can also visualize an existing database and what information is stored, create different views to visualize your database items, add data manually to your database to run some tests, and much more. Inside the **Data** tab, you will find other tabs to control privacy settings by determining who can access and modify this data. The **File manager** and **Option sets** tabs offer alternatives to store static options in a database-like structure that doesn't use the database, meaning it can be faster for some specific use cases.

The database of your app is a very important part of your application. If the workflows are the brain, the database is the heart, and they interact with each other very often.

Styles

In the **Styles** tab, you will find a list of all the components available inside Bubble, including buttons, texts, and inputs. If you are a designer, this is where you will customize the look and feel of each component inside your application, as if you were defining the visual foundation of your app. This concept is similar to a *design system* if you know the term.

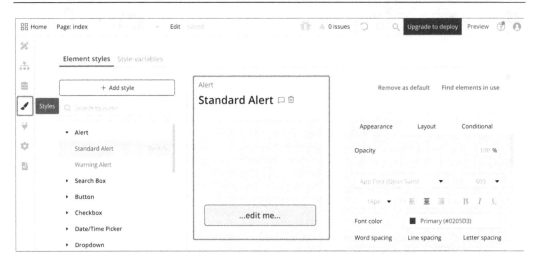

Figure 1.24: Bubble.io – Styles tab

Different from the **Design** tab, the **Styles** tab lets you define the visual characteristics of each element on a global level, meaning you can define specific styles for specific components and add a name to that style. This allows you to create a set of pre-defined styles for each component and select these styles while building your app's interface. The cool thing about using styles is that you can create multiple variants and organize, search, and manage your styles, all from the **Styles** tab. If you decide to change a style, it will alter every component attached to that same style name, meaning you don't have to go through each individual page and change all the styles one by one. This saves a lot of time when designing. Of course, you can also decide to change specific components and have them not follow a global style, which can be done in the **Properties** panel when using the **Design** tab.

Styles also allow you to add variables and conditionals for components, and these are also defined on a global level. As a designer myself, I really recommend using styles and taking some time to define the basic components of your Bubble app, even before the building phase, so you have everything set to get started and save time in the future.

So, to do a quick recap, you use styles to maintain visual consistency throughout your app by defining global properties such as colors, fonts, and sizes. By creating and applying styles, you ensure a unified design language across different elements, providing a cohesive and professional appearance for your application.

Plugins

Plugins are your superpowers, your magic tricks! The **Plugins** tab is the place where you add and manage your plugins. This tab displays all the plugins you have installed in your application and allows you to configure them. With plugins, you can enhance your app's capabilities by introducing extra functionalities built by other developers to your app. You can access the marketplace from the **Plugins** tab, where you can search and filter to find tools and resources that can help you build inside Bubble. There are thousands of options available, both free and paid for, that allow you to add extra functionalities to your projects, including payment gateways, API connections, new components, integrations with third-party services, and much more.

Figure 1.25: Bubble.io – Plugins tab

Now, if you want to go even further, you can. Know that you can also create your own custom plugins if needed, and add your own plugin to the Bubble marketplace. It can be built and used just by you (privately), but it could also be published publicly for other people to use, for free or paid for. Building plugins can also be a way to make money with no-code tools, but that is another topic. Let's continue.

Settings

The **Settings** area provides a central hub to configure your app's global properties. It allows you to set and visualize fundamental information about your Bubble web app.

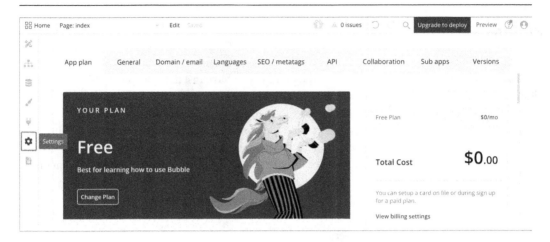

Figure 1.26: Bubble.io – Settings tab

Under **Settings**, you can manage your app plan and visualize the workload units used and your general privacy and security rules. You can also configure your custom domain and email settings, define your app's languages, configure SEO and meta tags, configure API rules and collaboration options, set up sub-apps, and see information about your Bubble app version.

This is where you can find important information and settings for your Bubble app and make sure everything is properly configured before launching, adjusting various parameters that influence your app's behavior to ensure that your application aligns with your intended goals.

Logs

The **Logs** section offers charts to help you identify how your application is performing. It is also where you can visualize and monitor your **Workload Units** (**WUs**) consumption. WUs are closely tied to the pricing plan selected by the app owner, and monitoring the consumption is very important as your application grows and experiences higher user engagement levels, as this is closely related to how much your app will cost to run. We will talk about that more in future chapters. For now, it is just important to know that you can track your workload usage visually, view app metrics and activity access server logs, and refer to the scheduler that works with backend workflows. The **Logs** tab can help during the identification and resolution of issues. By tracking errors, warnings, and app activities, you gain insights into how your app is functioning, facilitating effective debugging and optimization processes.

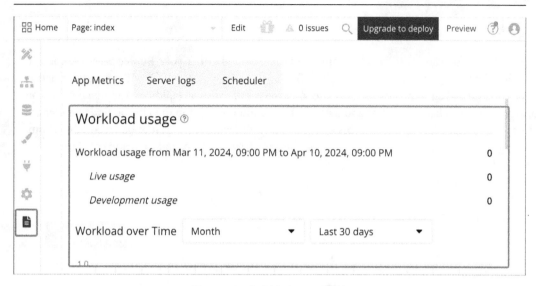

Figure 1.27: Bubble.io – Logs tab

At this point, you have a full picture of the main areas inside your Bubble editor and where you can find key features that will allow you to build and manage your web application. This is a great foundation to get started as you can now easily locate things and know where to go for specific actions and settings. Let's next dive into your account, which is the area outside of your editor, so you can continue exploring Bubble and how it works.

A quick tour of the main areas of your account

In the previous section, you learned about the Bubble editor and the main areas of each tab available inside the tool. All of that lives inside the **Builder** and will be used while you are working on your web app. Bubble also has a dedicated part to manage your projects and handle your account. In this section, you will learn more about the other side of Bubble, which is not related to your project specifically but to your account. Let's dive into the sections and areas you will use while not building with no code.

Apps

Under **Apps**, you can visualize all the apps you have created under your Bubble account. You can search and filter apps if you have a lot of them. By clicking an app card, you can view more information about the app and choose to preview it or go to the editor to continue building. Additionally, you can also add an icon, change its color and name, and even pin it to the top to help you find and identify the app better. You can also check this app's current plan, when it was last updated, and by whom, add tags, share the link with others, duplicate the app, delete the app, and even add collaborators to work with you on this project.

Figure 1.28: Bubble.io – account apps

If you click on the application item, you can view more information about it and choose to launch the editor, preview the app, share the link, and much more.

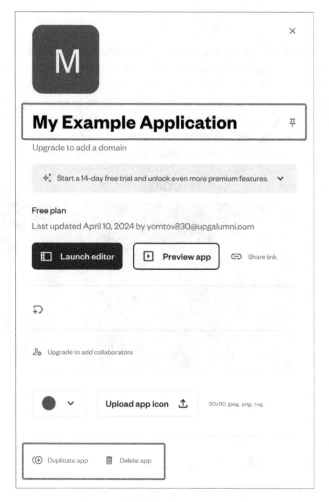

Figure 1.29: Bubble.io – app details

You can also customize your app name and color and upload an icon so it is easier to identify which app is which. This is particularly helpful when you have multiple applications created under the same account. If you have too many apps, you can also choose to pin this app so it will be displayed at the top of your apps list, meaning it will be easier to find it. You also have access to a search bar and filters in case you need to find an app that is not visible on the first page of the apps section.

If you need to store information about an app for your team members or yourself in the future, you can also add notes to it. It is also possible to work with other people on the same project and invite collaborators to work with you on a particular project. In this section, you also have options to duplicate the app, creating an exact copy for various reasons, and if you don't need the project anymore, you can also delete it forever and clean up your account a little bit.

Account

You can access your account at the top-right corner of your window by clicking your avatar image. Under your account, you can change and check your billing information, change your account password or email, log out, reset your password, and check your security options to make your account more secure by adding 2FA, for instance. You can add your own custom avatar image, your first and last name, and even apply to become a Bubble developer. You can visualize credits and for how long you have been bubbling. Under **Billing**, you can see your current plan, add or remove your card and billing information, and enter coupon codes. Under **Invoices**, you can visualize your plans and invoices.

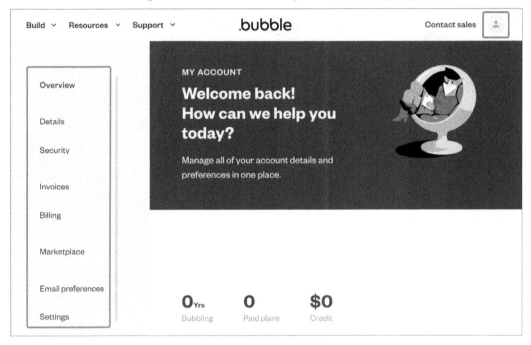

Figure 1.30: Bubble.io – account overview

Templates marketplace

Bubble has a templates marketplace, where you can search for and filter amazing templates and design components, both free and paid-for, built by other bubble developers to help you build faster. You can choose to preview and interact with a template to test and see how it works. If you decide to use it, you can click a button to start a new project and clone that exact template. Sometimes, templates are simple, single pages or specific design components that help you build faster. Other times, they're entire working projects that can go from beginner to advanced level. You can also choose to create your own templates and either add them to the marketplace for free or sell them. If you wish to sell a template, Bubble will take a commission for every sale. To add a new template, create a project and then turn it into a template. You can also check the **My Templates** item under your account.

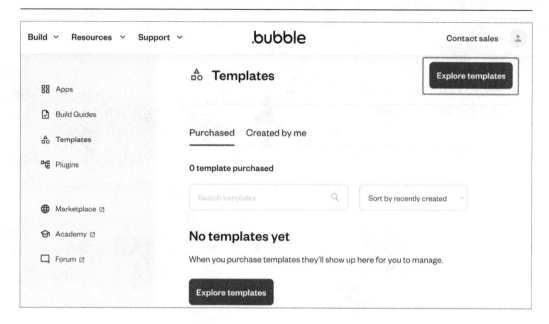

Figure 1.31: Bubble.io – Account templates

Using templates can be a good idea when starting a new project to save time and speed up your building process. Just be careful when using templates, because some can be challenging to understand and work with, so consider your existing level of Bubble experience before jumping into a too-complex template and getting stuck.

Inside the templates marketplace, you can search for diverse options and also use filters to help you find the specific template you are looking for. You can find a lot of free templates to use and also premium paid-for templates. It is fun and recommended to explore templates and see how they were designed and built inside the editor, this can also allow you to learn how to use Bubble. You should give it a try.

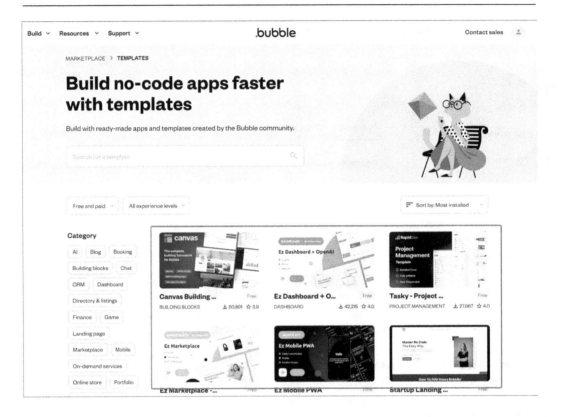

Figure 1.32: Bubble.io – Templates marketplace

Plugins marketplace

From your user account, you can access your **Plugins** area and see the plugins you have added or created. If want to find plugins, you can go to the Bubble marketplace via the website, or from inside a project, you can find plugins under the **Plugins** tab. I strongly recommend getting familiar with plugins because they allow you to build amazing things by leveraging extra features and third-party services that integrate with Bubble.

The **Plugins** page inside your Bubble user account is shown in the following screenshot:

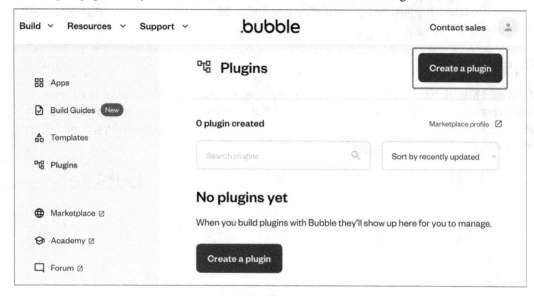

Figure 1.33: Bubble.io – Account plugins

In the plugins marketplace, you can find thousands of plugins ready for you to plug and play into your Bubble web apps. Using plugins is a very good way to add powerful functionalities to your builds.

The plugins marketplace page on the Bubble's website in the following screenshot:

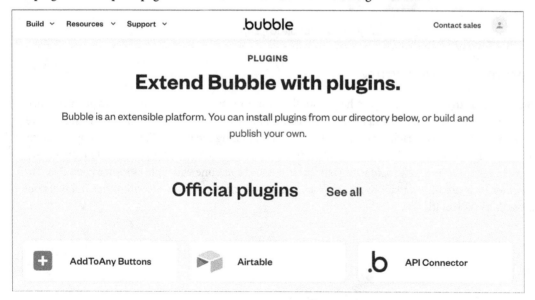

Figure 1.34: Bubble.io – Plugins marketplace

With plugins you can go beyond what is possible with just the Bubble editor and native features, so make sure to take some time to explore plugins and find good ones that can take your project build to the next level.

Support

If you need help, Bubble offers support, especially if you are on paid plans. You can send a message via their website or send an email to open a ticket. To do that, log in to your account and find the **Support** tab. From there, you can contact sales, go to the support center, talk about partnerships, report a bug, report abuse, or just reach out to get help.

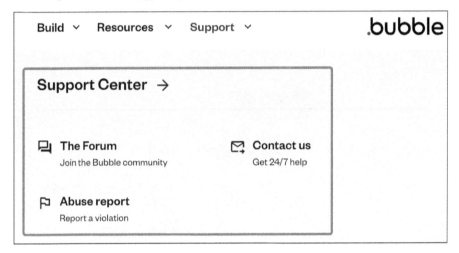

Figure 1.35: Bubble.io – Support Center

Forum and community

Bubble has a strong community, so help from the support team is not the only way to get assistance. There is a forum available for you to join and ask questions. The Bubble community is very supportive, so if you want to learn, teach, or get help, I recommend joining the forum. You can start by searching forum posts to check whether someone already asked your question, and if you don't find an answer, you can easily create a new topic with your questions and someone will reply as soon as possible. The community is not only a place to get support, but to also meet and connect with other Bubblers out there. Make use of it!

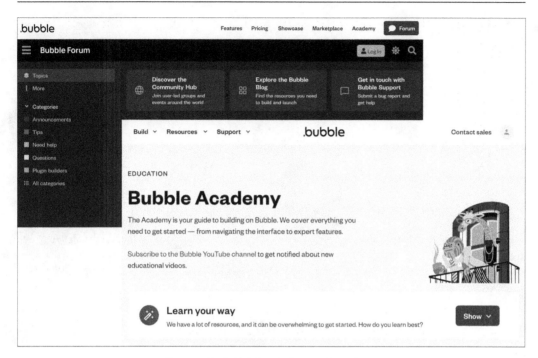

Figure 1.36: Bubble.io – Forum and Academy

By now you know what Bubble is, what you can build with it, and how to navigate the tool from the inside out. This knowledge is helpful because one of the main challenges for beginners when using a new tool is to identify where things are, and now you know every section of the Bubble editor and your account, so you will be able to navigate freely and locate features as you build your applications.

Summary

In this chapter, we've walked through the foundation of your Bubble learning journey by teaching you the first steps to be able to use this powerful tool. We covered the essentials about Bubble, what it is, and what you can build with it. You also learned how to create an account from scratch and how to set up your project using the application assistant.

We covered every area of the main editor in detail and saw how to navigate inside Bubble with a quick tour of the main areas of your account, so you know how to find anything you need while building and managing your applications.

This step is very important to guide you on your path to using Bubble and will form a strong base to allow you to understand more complex concepts in the future. In the next chapter, we will dive into the components and learn in more detail what they are and how they work.

2

Navigating the UI Builder Components Tab

The process of building a web app consists of adding components to a blank canvas (the editor) using drag-and-drop features and making sure you have the right building blocks added to the page to build the perfect layout to achieve your desired goals.

In the previous chapter, you learned how to create your account and what are the main areas of the Bubble editor. In this chapter, you will take a closer look at the **UI Builder** components tab available inside Bubble. You are going to learn the key areas and features, what they are, and how they work.

This knowledge is valuable because it will allow you to understand the core features of Bubble and how to use and interact with the UI sidebar and its components, allowing you to build your own applications with ease.

This chapter will cover the following topics:

- Presenting the **UI Builder** components tab and the sidebar
- Understanding the structure of the **UI Builder** tab
- Learning about the **Elements Tree** and how it works (visibility and conditional settings)
- Exploring important concepts such as container components, input forms, and the power of reusable elements.
- Learning how to install additional components from the Plugins Marketplace
- Gaining insights into recommended practices and tips for efficient component management and utilization

Presenting the UI Builder components tab and the sidebar

Let's break down the components by category so that you can follow the Bubble editor as it is from top to bottom. We are going to start on the **Design** tab, which is the main tab you see selected when joining your Bubble editor. We are going to start with the sidebar items, making sure the **UI Builder** tab is selected at the top of the **Visual Elements** sidebar.

The **Design** tab and sidebar **UI Builder** tab are selected, as shown here:

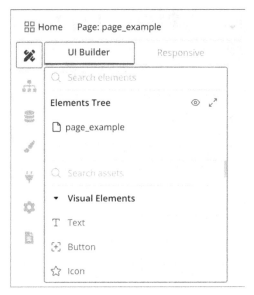

Figure 2.1: Design tab – UI Builder

Before we talk about UI components, it is important that we understand what the Elements Tree is; this is the first area on your component's sidebar, as shown here:

Figure 2.2: UI Builder – Elements Tree

Under the **Elements Tree** section, you will see each component that was applied to your page, meaning if you drag a new component to the canvas, it will show up there.

The UI Builder Elements Tree with **Title** and **Button** components is added to the canvas, as shown here:

Figure 2.3: Elements Tree example components

This concept is similar to other design tools such as Figma and Photoshop; these are your "layers," but in this case, for your components, the elements tree will show all items that are present on your page, even hidden ones.

Sometimes, you will want to hide a component that is present on your page, but just visually, because maybe you want to see other components that are below existing components or just because hiding it for a while helps you do something inside the visual editor.

So, if that is the case, there is a quick way to identify whether a component is visible or hidden or to make it visible or hidden. Check if there is a little eye icon on the right side of the component name and if this eye is crossed or not crossed. If it is not crossed, it means the component is visible; if it is crossed, it means the component is applied to the page, but it is invisible. You can also click the eye icon to toggle it and choose if you want to make that component visible or invisible.

The UI Builder Elements Tree with a hidden component is shown here:

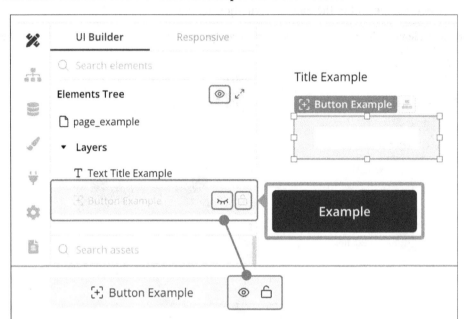

Figure 2.4: Elements Tree eye icon

Important note

 When you show or hide a component using the eye icon, it doesn't affect whether the component applied on your page will be visible to your users. It only changes the state of that component on your Bubble editor. There is a difference between what you as a builder see inside the editor and what your users will see when you let them navigate on your application, so keep in mind that changing the eye icon doesn't show or hide the component when the application is loaded; it only affects the editor view, where you are working. If you want to hide a component when the page loads, there is an option inside the property editor for that. We are going to cover that in the next sections of the book.

When using container components, if there are groups or components that are inside another component applied to your page, you will see a plus icon next to the component name, meaning you can click here to expand the component and see other components inside it. This is what we call a components tree, meaning the first component is the parent component and the components inside it are their children.

The UI Builder Elements Tree with a **Group** component applied to the canvas is shown here:

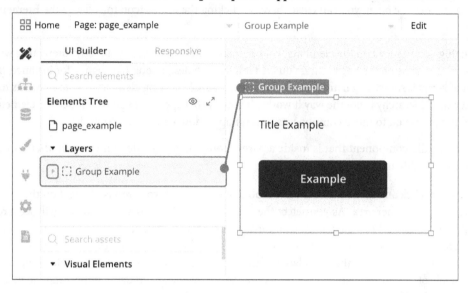

Figure 2.5: Elements Tree container components

An example of a container component expanded showing other components (children) that are inside this (parent) component is shown here:

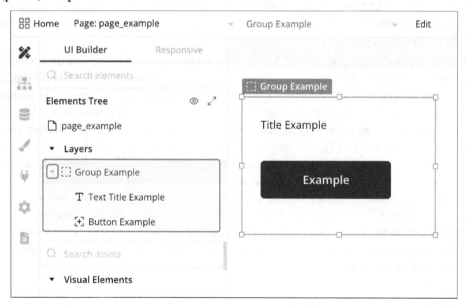

Figure 2.6: Elements Tree container components opened

Container elements are shown with a plus icon at the left to allow you to open them and see what is inside. Once they are open, you can close them by clicking the minus icon to collapse the **Elements Tree** list.

Rearranging components on the elements tree can also be done, but not by directly interacting with the component name on the tree itself. If you are familiar with design tools like Figma, for instance, the concept of layers is something natural to you, but inside Bubble moving elements on the elements tree to change their hierarchy is not the way it works. If you try that approach you will notice the elements tree will not allow you to move components (layers) up or down, like other tools do.

To change a child component that is inside a parent component outside of it or to another parent, you can use two different approaches:

1. Visually click and select the desired component (child) with your mouse, drag it on the editor canvas to another area. As you select the component and start dragging it, you will see a red line indicating who is the parent component. For instance, you have a text component inside a group, that group, which is the parent, will become highlighted in red, indicating the text is inside it, and where are the boundaries of that component. When you drag the child component outside of the current parent, another red line will show up, on another parent component. This means that if you drop the component on that area of the screen, that highlighted component in red will become the new parent container of the component you are moving. Give it a try! After you've done that, notice that the elements tree order will change, the component you moved on the page will change order on the list and will be placed inside or outside of a new group, depending on your changes. Note that this method works but can be a little bit harder to control depending on the complexity of your layout, which leads to the second approach, which you might prefer.

2. Use your keyboard shortcuts to CUT the element from the page. This will remove the component completely from the current parent and allow you to move it somewhere else. Use the shortcut keys: Windows (ctrl+x), Mac (cmnd+x). Once you do that, your component will be removed from your page, don't worry it is not lost, it is stored on your clipboard. Now, click to select the new parent component you want this cut component to go into and then with the new parent component selected, PASTE the component inside it. Use the keyboard shortcuts: Windows (ctrl+v), Mac (cmnd+v). Done, your component is now placed inside a new parent. Note that you can also use the "Edit" features available on the Bubble top bar, it will work the same way. I find it easier to use shortcuts to make it faster, but feel free to use what you prefer. If things go wrong, don't worry, you can always go back. Note that to move one component to another parent, usually the parent components must be groups or similar types of components that allow other components to be placed inside it. If you select a text for instance and then try to move another text inside it, it won't work.

Inside the **Elements Tree** section, there is also an option called **Only show hideable**:

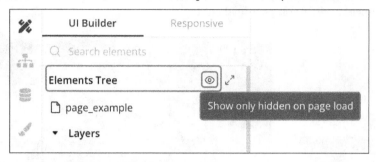

Figure 2.7: Elements Tree – the Only show hideable option

This option is turned off by default; if you click the checkbox on the right side and mark it as active, it will filter the components that are on your page that are only loaded with certain conditionals. This means it will show components that exist on your page but that are hidden by default, meaning these components are not loaded when the page loads. If you have components that are configured like that, once you check this option, it will expose which components are in that specific condition and hide others so that you can easily identify them inside your page editor.

Each component inside Bubble has an option to make the component hidden by default, meaning when you first open a page, if that condition is checked, that component will not load with the page load and be displayed on the page. But the key thing is, that component will still be visible on your Bubble editor, so that is why this option exists: to help you identify components configured this way.

As explained earlier, that option doesn't have to do with the eye icon, so it is not about the component being hidden using the eye icon.

If you click **Only show hideable** but there is no component configured to not load on the page load, it will just hide *normal* components on the **Elements Tree** list and show an empty list. If there are any, it will expose which component is under that condition and hide the other ones. The best way to understand this concept is by trying it yourself, so I advise you to play with the tool and get familiar with it.

As additional information, here is where you set up this option to not show the component on the page load.

Select the desired component and go to the **Properties Editor,** click to expose the **Layout** tab, and find the **This element is visible on page load** option:

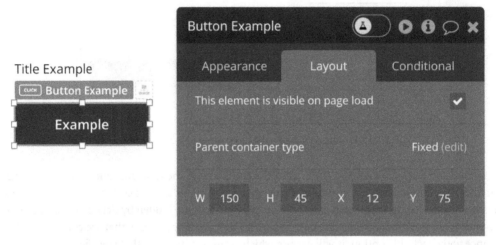

Figure 2.8: Property Editor Layout tab – element visibility

Check or uncheck this option to define whether the component loads on the page load. The default is that every component will load, so it is automatically checked; if you don't want it to load, uncheck it.

Quick tips about the Elements Tree

If you click on top of the component name, it will be highlighted on the page, bringing you to the position where the component is. That can be helpful to locate and select specific components.

The first component name is the page itself, so if you want to target the page, just click the **Page [Name]** item on the elements tree.

Below the elements tree, we will also find a search bar. This search bar can be used to help us find a specific component we are looking for. This means we can use it to locate a specific UI component by typing its name; for instance: Repeating Group. Once you type something, it will filter the components and only show the ones matching your search keyword. This can be helpful when you need to locate components, especially if you installed new ones and you don't remember their exact name.

Note that this search is not to find components applied to your page; there is another search at the top bar for that. This search is to find components inside that same sidebar, such as text, buttons, icons, and so on. It is just a way to get to the component you are looking for faster:

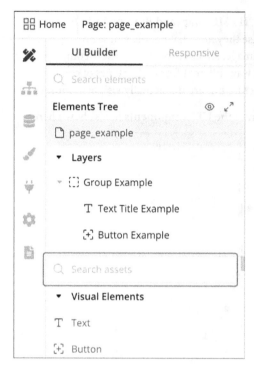

Figure 2.9: UI Builder search bar

The elements tree is a great place to find and locate your components in case you have multiple things added to the page and hidden parts that will only be visible during certain types of user interaction with the page and conditionals, so remember to check this area to find your components and easily locate them.

Presenting UI components

You learned all about the Elements Tree; now, let's dive into UI components. It is important to notice that the visual components are divided by categories, inside the Bubble editor. Let's break down each section so that you can learn what they are and what types of components you will be able to find in each category.

Visual Elements

These UI elements, as the name of the category suggests, are visual little blocks you are going to use to build pretty much any application. We can call this section the basic elements that will help you build web apps; it is very unlikely that you are not going to use a text component or a button on your application. Let's say these are the most used and essential components for any project you are going to build using Bubble. Inside the **Visual Elements** category, you will find **Text**, **Button**, **Icon**, **Link**,

Image, **Shape**, **Alert**, **Video**, **HTML**, **Map**, and **Built on Bubble** elements. These UI components are called visual elements because they are what people will see on a page; for instance, a text with a **call to action** (**CTA**) button, an image, an icon, and so on. These components are what will compose your application layout in a visual way, different from other types of components, such as containers, which serve to create the page structure but are not actually visible in most cases.

The **Visual Elements** section of the **UI Components** list is shown here:

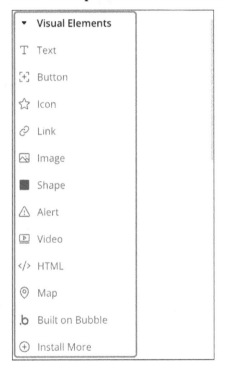

Figure 2.10: Visual Elements section

Don't worry – we are going to learn more about the other types of components in the next section, so let's continue.

Containers

Under **Containers**, you will find elements to build the foundation of your layout and page structure. Imagine you are building a house; well, these components are the walls that hold everything together, dividing rooms and creating spaces to organize other components. In this category, you will find **Group**, **Repeating Group**, **Popup**, **Floating Group**, **Group Focus**, and **Table** elements. Containers act as frames or groupings for other elements, meaning you will potentially add visual elements and other components inside a container such as a **Group** or **Repeating Group** container. This will allow you to control the arrangement and presentation of content within your app.

The **Containers** section of the **UI Components** list is shown here:

Figure 2.11: Container section

Whether it's grouping elements together or creating responsive designs, containers play a pivotal role in achieving a cohesive layout structure, so make sure to understand how they work, and play with containers to get familiar with them.

Input forms

These elements are responsible for collecting and capturing data from your users, usually used to build forms and dynamic features that will interact with the database of your application. For instance, with inputs, you can create a form for a contact page or a signup and login page. Inside this category, you will find **Input**, **Multiline Input**, **Checkbox**, **Dropdown**, **Searchbox**, **Radio Buttons**, **Slider Input**, **Date/Time Picker**, **Picture Uploader**, and **File Uploader** elements.

Inside Bubble, you can customize forms with flexibility, choosing from any of the options available to collect data from your users. You can add fields such as text inputs, checkboxes, drop-down menus, and much more.

The **Input forms** section of the **UI Components** list is shown here:

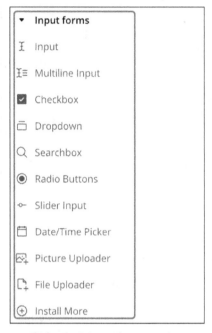

Figure 2.12: Input forms section

Input forms are crucial for collecting user information, conducting surveys, and enabling data submission, making them a fundamental part of many applications.

Reusable elements

With **Reusable elements**, you can save time and optimize your app-building process inside Bubble, because with reusable components, you can change components one time and have them changed across the entire application. You can easily turn a normal component into a reusable component or even group a few components (for instance, an entire section) and make it a reusable component.

The **Reusable elements** section of the **UI Components** list is shown here:

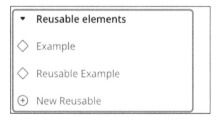

Figure 2.13: Reusable elements section

Reusable components work in a very specific way. Once a component is turned into a reusable component, it no longer can be edited via the page (canvas) editor. It will gain its own place and "page" to be edited; this page can be accessed via the **Pages** area in the top bar of your Bubble application, under **Reusable elements**.

The top bar, **Pages** list, and **Reusable elements** list are shown here:

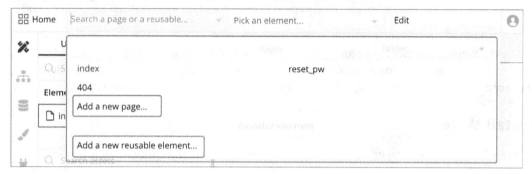

Figure 2.14: Reusable elements page menu

Once you add a reusable element to the page, it will act like a clone of a mirror. If you add it to multiple pages, it will look and function the same way on all the pages, and if you make a change to the original reusable component, this change will be applied everywhere.

Here's an example of a reusable element in the editor canvas:

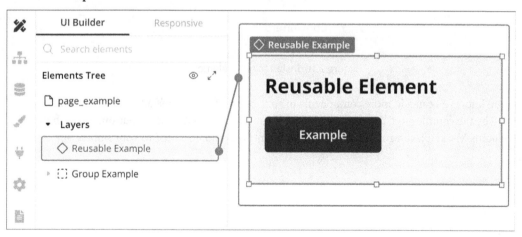

Figure 2.15: Reusable element added to the page

So, for instance, if you have a header on a website, and you want to update a link, you only need to do it once if using a header that is a reusable component. All pages that have this header component added to the page will be automatically updated following the changes you did inside the reusable component. It is like the reusable component is the master component and all the pages have a copy of that exact component, so if it changes in the master, all the "clones" are updated as well.

This is a very cool and useful feature to help speed up your building process, keep consistency, and maintain projects with quick updates.

Under the **Reusable elements** category, you will see all available reusable elements you created with their respective names. To use a reusable component, just drag it to the page, and it will be applied. Further in this book, you are going to learn how to create reusable components step by step.

Install More

In every UI component category, you will find a link saying **Install More** with a plus icon.

You can see the link here:

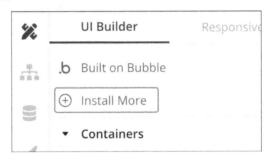

Figure 2.16: Install More UI element

Did you know you can add more components to your Bubble editor by installing external components created by the community? Yes, that is right. So, if you click the **Install More** button, it will show you the Plugins Marketplace window filtering components that can be added to your Bubble app.

The **Install new plugins** modal window is shown here:

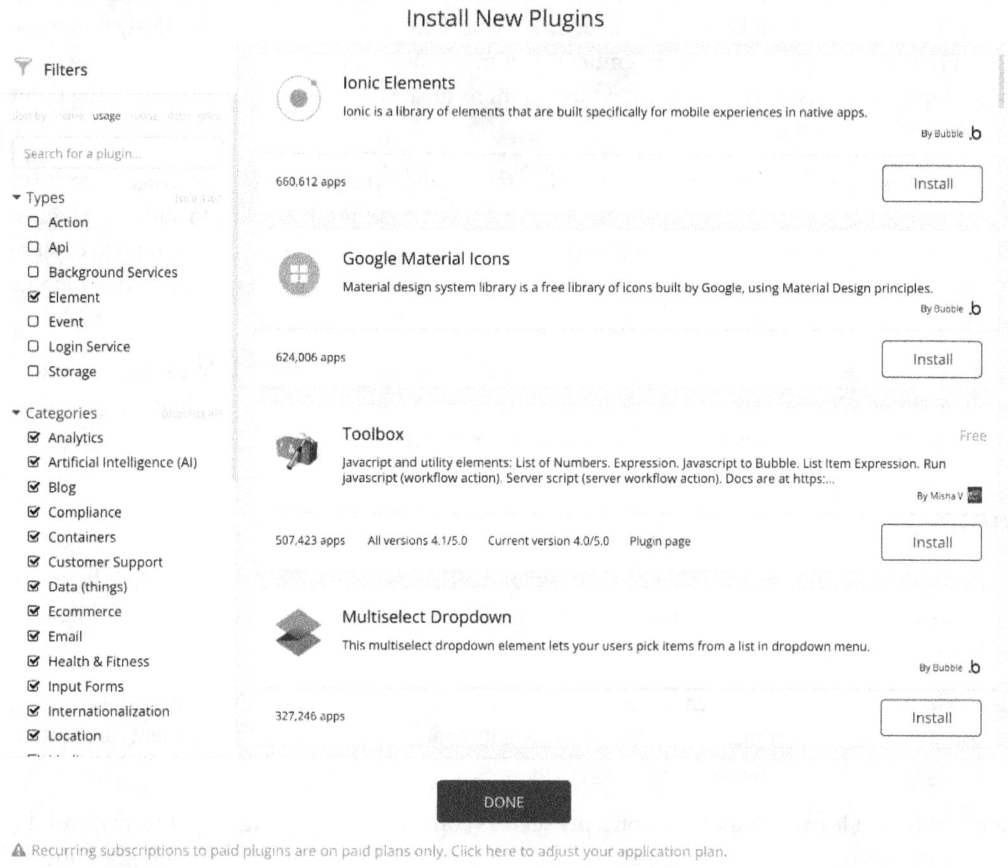

Install New Plugins

Figure 2.17: Plugins Marketplace

In this area, you can search and find new components that can help you build your application; some are free and others are paid. This is a cool way to find even more powerful components to build awesome applications inside Bubble. But also be careful to not add a lot of new components that can make your application heavy.

A good thing to know is that the new components are in fact plugins, so that I why you are going to see the Plugins Marketplace. The only difference is that some plugins are components that will be applied to the frontend, and others are going to be applied to the backend or both. There are different plugins available in different categories inside the Plugins Marketplace.

Sometimes when you install a plugin, that plugin will require you to add something on the frontend of your application to make the plugin work. If that is the case, the plugin component will show up as a new UI component available somewhere inside your **UI Components** list, most likely under the **Visual Elements** category. So, just remember if you install new visual components, it is going to be visible under the **Visual Elements** section. Just go there, locate the new items added, and drag them to the desired page to start using them.

Adding plugins is a very nice way to extend Bubble; for instance, you can install new icons, a library of buttons, new form components, mobile-compatible UI kits, and so on. The ability to add new plugins, components, and new resources to Bubble makes it even more powerful, so take some time to explore available components and plugins. In future chapters, I will give you a list of recommended plugins and components you can use!

Now that you know all the UI Builder categories and the components available in each section, let's talk about each individual component in more detail so that you can know what they are, how they work, and what you can do with them. See you in the next chapter.

Summary

In this chapter, we explored UI Builder components and the sidebar, getting familiar with their structure and functionality within Bubble. This foundational understanding helps you navigate Bubble's interface effectively while working on your applications.

You also learned about the Elements Tree and how it can help you control component visibility, along with using conditional settings efficiently. These insights enable you to manage components accurately, enhancing your application's usability.

Additionally, you learned important concepts such as container components, input forms, and the benefits of reusable elements. You also discovered how to expand your toolkit by installing additional components from the Plugins Marketplace, opening up new possibilities for your application development journey.

Throughout the chapter, you gained practical insights into best practices and efficiency tips for managing and utilizing components effectively. With this knowledge, you're ready to explore UI components further. In the next chapter, we are going to dive deeper into the UI components available inside Bubble.

Building Blocks – Exploring Bubble's UI Components

In this chapter, we are going to continue learning about **Bubble** and **user interface** (**UI**) elements. We are going to dive deeper into the tool and each element category, so you can get familiar with all the existing components that can be used to build your page layouts. You will learn not only what the elements are but also what they do and what they can be used for.

This knowledge will be valuable to building any project using Bubble as you will know what exact components to use when building your own digital applications.

Let's dive right in and start learning about visual UI elements inside Bubble. Here is what we will cover in this chapter:

- **Presenting visual elements**: Text, Button, Icon, Link, Image, Shape, Alert, Video, HTML, and Map

- **Presenting containers**: Group, RepeatingGroup, Popup, FloatingGroup, and GroupFocus

- **Presenting form elements**: Input, MultilineInput, Checkbox, Dropdown, SearchBox, RadioButtons, Slider, Date/TimePicker, PictureUploader, and FileUploader

- Explaining what reusable elements are, how to create them, and how they work

- Explaining how to install and use new design components

Diving deep into UI elements

At this point, you have learned where to find the components inside Bubble and how they are divided by category. Now, it is time to dive deeper and understand more about each one of them, so during your development process, you will already know what to use for each project you want to build. Let's dive right in.

Visual elements

The UI elements available inside Bubble are divided into categories – the first one is called visual elements. These are the most used components to add visual content to your pages, so let's learn a bit more about each one of them.

Text

With this component, you can add text to your page and play with the typography. You can draw a box and type the desired text on the property editor to add your copy. It is also possible to add dynamic data to the Text field, meaning the Text information can come from a database source.

The **Bubble Editor** window with `Text` elements on the page is shown here:

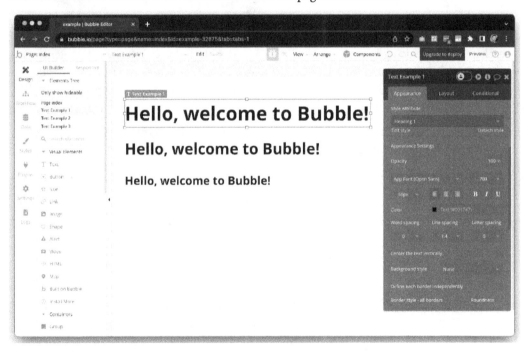

Figure 3.1: Text component

The `Text` component also allows you to play with a rich text editor, and it also can recognize links and emails.

You can choose whether the content will but cut off if the area you drew is not big enough or shrunk, in case there is less text than the area you designed.

Inside the property editor, on the **Appearance** tab, you can define the font family, weight, size, color, and opacity, define whether the text should be centralized or aligned left or right, make it bold, italic, or underlined, select the word spacing, line spacing, and letter spacing; you can also select an existing text style and much more.

Button

The Button component is very important, as it will allow you to perform various actions, such as submitting forms, building navigation adding a **call to action** (**CTA**), triggering workflows, or navigating to different pages within your app.

The **Bubble Editor** window with a Button component on the page, as shown here:

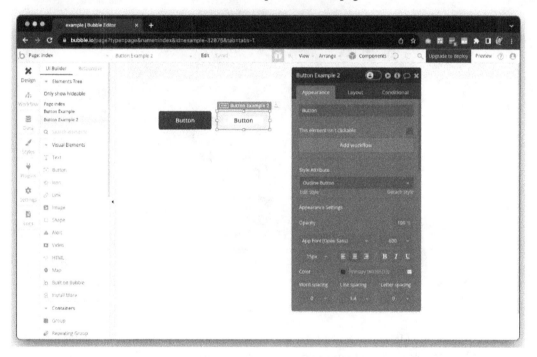

Figure 3.2: Button

You can customize buttons by defining their text, appearance, and behavior. Your Button labels can be static or dynamic, you can adjust their size and shape, round corners, choose colors, and define hover and animation effects. Buttons can also be responsive and adapt to your layouts.

Icon

The `Icon` component allows you to add visual information to your layouts and make your designs more interesting also by conveying information quickly and intuitively. By default, this component comes with a library full of icons for you to choose from; you just add an icon to the page and then select from a list which icon you want to display, and the icon will fit its container and be of the size you draw it.

The **Bubble Editor** page with an `Icon` component is shown here:

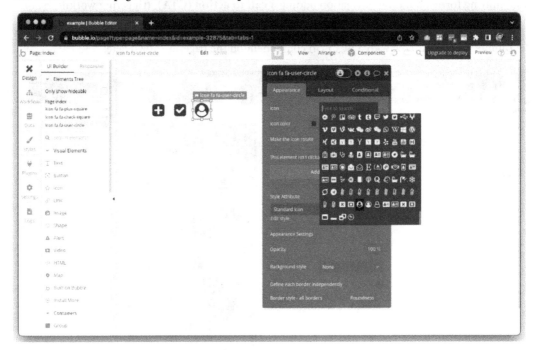

Figure 3.3: Icon

The library of icons Bubble uses is called Font Awesome, which is a free open-source icons library, so you don't have to pay anything to use icons or add icons yourself. Now, of course, if you want to add your own custom icons, you can do it, but then you are not going to use the icons element for that – you are probably using images or other components to hold your custom icons. Another way to add custom icons is by searching the marketplace and installing new libraries. With the native Icon component, you can customize their size, color, and alignment to fit seamlessly into your app's UI. Icons are particularly useful for representing actions and categories or adding visual appeal to your app's design.

Link

Different from the `Button` component, a `Link` component is like a `Text` component but with extra features and settings to define. You can, of course, add a link to a `Text` component, but if your goal is to add links, then this component is more specific to that.

The **Bubble Editor** page with a `Link` component is shown here:

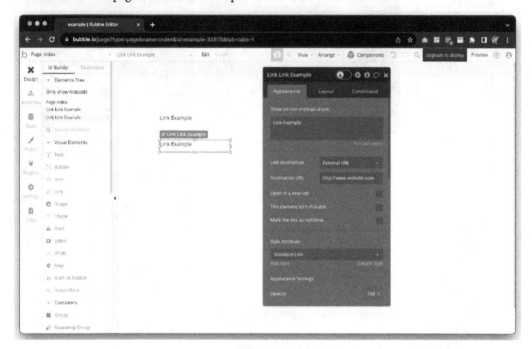

Figure 3.4: Link

With this component, you can create hyperlinks, allowing users to navigate to different pages of your app, other websites, or external resources. You can define the link's destination URL and even send customized data from one page to another. You can also customize its appearance, including text, which can be static or dynamic, change the font and color, make it change colors when hovered or clicked, and add a custom style to make all links look and behave the same way. Links are essential for providing seamless navigation and enhancing the user experience by making content easily accessible.

Image

With the `Image` component, you can add photos and graphics to your application. Images will play a huge part in your layout design. You could also use images to add custom icons if needed. Once you add an `Image` component to the page, you can upload images directly to the component, bring images from the database or an external API, or reference external image URLs; the component is very flexible.

The **Bubble Editor** page with an `Image` component on the page is shown here:

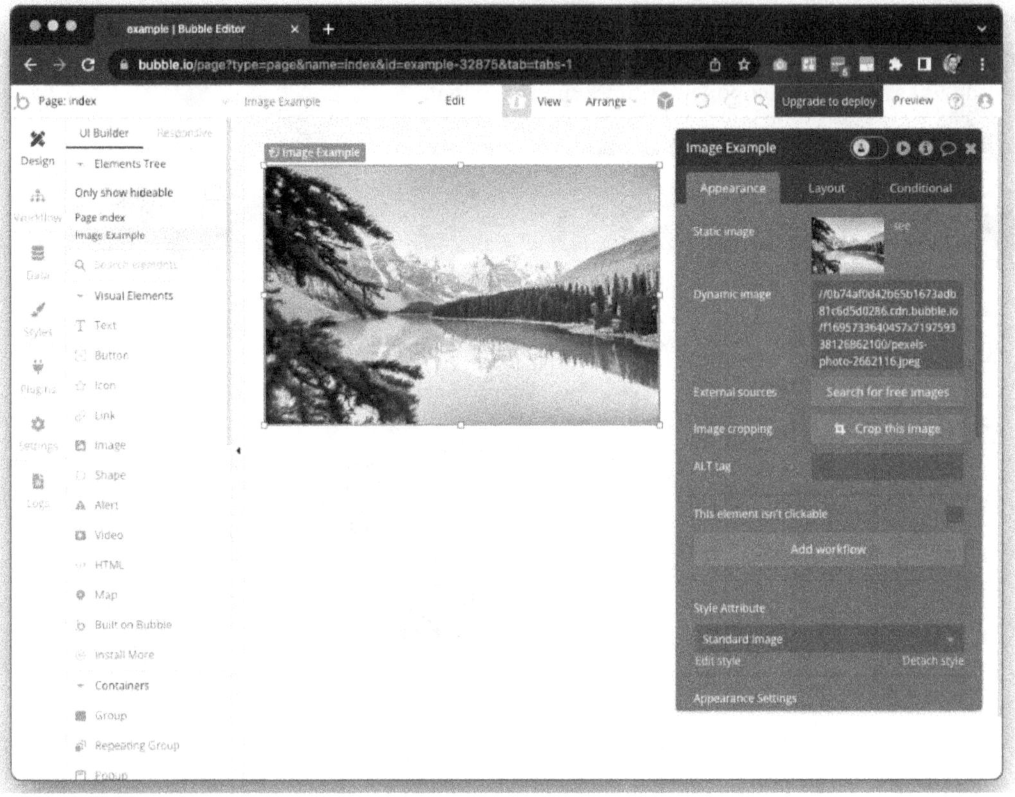

Figure 3.5: Image

You can customize image dimensions, apply various styling options, such as borders and shadows, and control how images are going to stretch or remain static to fit your layouts and make them responsive while working inside containers and other components. Images are fundamental for presenting visual content, such as photos, logos, and illustrations, and can help your app's design to delight your users. Bubble also has a feature to help you find images – if you click the **Search for free images** button under the properties editor while selecting an `Image` component, it will show you an image database you can use to search and choose images to add to your component. Simply find one you like and click it to add the selected image to your `Image` component; the image will be automatically downloaded to your project and will be ready to use inside your image component.

Shape

With this component, you can add rectangular shapes and decorative elements to your pages. You can draw inside sections to add, for instance, a background color, a square or rectangle element, or a divider and customize the shape dimensions, colors, borders, and opacity to align with your app's visual style.

The **Bubble Editor** page showing a `Shape` component is shown here:

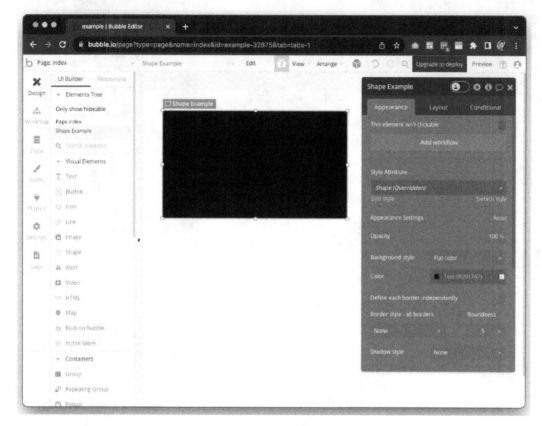

Figure 3.6: Shape

This is a basic component that can be used in many situations. A quick tip is to not use shapes as containers and to use containers as shapes for a more effective layout structure.

Alert

By default, the `Alert` component, once added to the page, is invisible, meaning when you load the page, it will not show up. This component serves as a notification tool to communicate important information to users. For instance, after filling out a form, you can use an alert to display a message to the user saying whether the form was submitted successfully or not. You can create custom alerts; inside the `Alert` component area, you can add and play with text, icons, and styling options such as colors, borders, and more to create alert messages for various scenarios such as warnings, errors, or confirmations. The Text elements can be static or dynamic, getting data from the database.

The **Bubble Editor** page with an `Alert` component on the page is shown here:

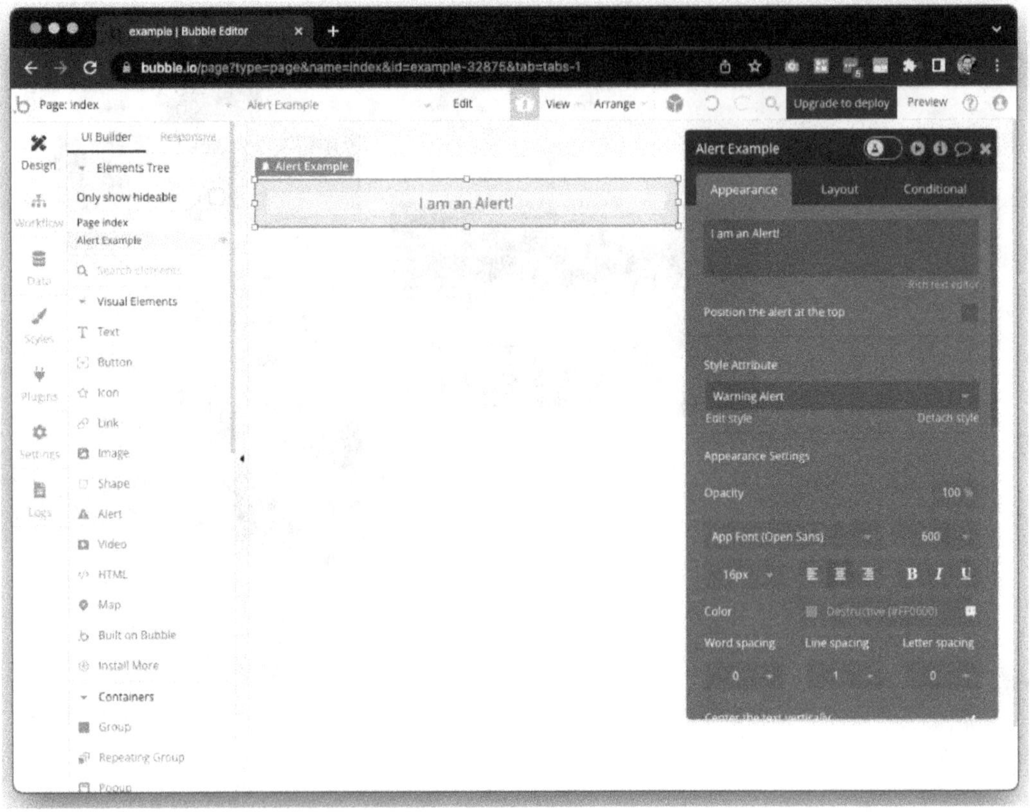

Figure 3.7: Alert

Remember to customize the Alert behavior and create specific workflow logic to make the alert show or hide. You can make it appear for a few seconds and hide automatically or add a button to dismiss the message – it is all flexible and customizable. You can also find custom `Alert` components made by the Bubble community at the marketplace in case you need something different. Alerts are essential for providing feedback to your users and play a huge part in the user experience, so, make sure to use them to enhance the usability of your app.

Video

With this component, you can add videos to your pages. Once you add a `Video` component, you can link to video files coming from your app's database or choose YouTube or Vimeo videos. It is possible to customize the video dimensions, controls, and autoplay settings.

The **Bubble Editor** page with a `Video` component on the page is shown here:

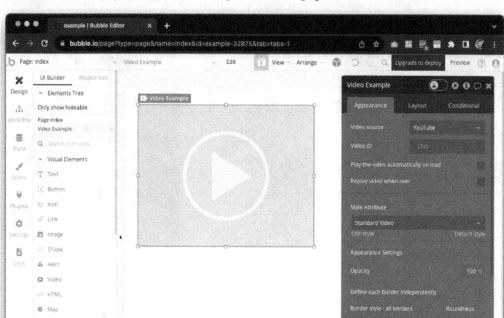

Figure 3.8: Video

Videos are effective for presenting multimedia content, tutorials, product demos, or any visual information that requires motion and sound. If the default `Video` component doesn't suit your needs, you can find more advanced `Video` components available in the marketplace and install them on Bubble.

HTML

Bubble is a no-code tool, but it allows you to add a little bit of code inside your pages if needed. There are many reasons why you would need to add HTML code to your Bubble page, for instance, when you need to add an embed or integrate external services via widgets or scripts.

The **Bubble Editor** page with an HTML component on the page is shown here:

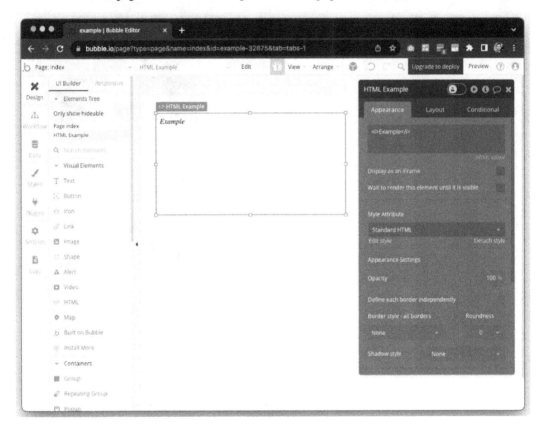

Figure 3.9: HTML

You can also choose to add a few lines of HTML, CSS, and JavaScript to customize the app's behavior or fine-tune your app's functionality beyond Bubble's native components. Remember that the HTML component added to the pages will load on page load and will only affect that specific page, not the entire app.

Map

Bubble allows you to add maps to your pages. You can drag or draw an area to show specific locations or pins on a map using the Google Maps API. It enables you to integrate dynamic maps into your app, providing location-based functionalities. You can display maps with markers, routes, and interactive features that let people click, zoom, and play with the maps. You can customize maps with various styles, zoom levels, and map sources.

The **Bubble Editor** page with a `Map` component on the page is shown here:

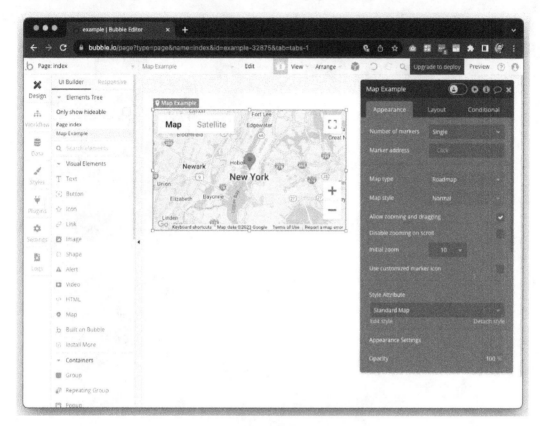

Figure 3.10: Map

Maps are essential for creating location-based apps, geolocation services, or displaying geographic data to users. Remember that using maps will rely on your Google Maps API key and this is a paid service provided by Google; using maps will have a cost, depending on the number of maps you add to your project, how many users are using them, and how frequently maps are loaded, so consider this cost if you plan to build an app based on maps. To know more about it, check out the Google Maps API website and their available plans.

BuiltonBubble

You can use this component if you are very proud of using Bubble – it will allow you to show the world that your application was built using Bubble. Usually, while using software on the free plans, you are forced to show some sort of stamp or logo to advertise to your audience what tool you are using.

The **Bubble Editor** page with an example of the `BuiltonBubble` component on the page is shown here:

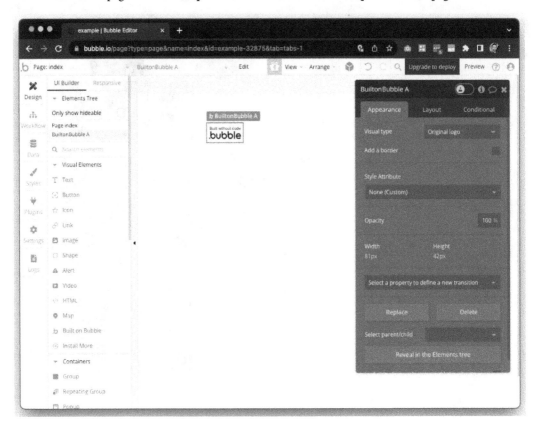

Figure 3.11: BuiltonBubble

You don't have to use this component, especially if you are on a paid plan, but it can be used after upgrading to a paid plan if you still want to show the world that your app was built using Bubble. Once you add it to the page, it will display a little badge that will let people know your app was built on Bubble.

Containers

This category contains elements that can be used to create your page layouts and structure. Consider containers as boxes that hold other components together – let's learn a bit more about each component available inside this category.

Group

With a Group component, you can create your page layouts and structure; this is a versatile element that lets you group multiple components, as the name suggests.

The **Bubble Editor** page with a Group component on the page is shown here:

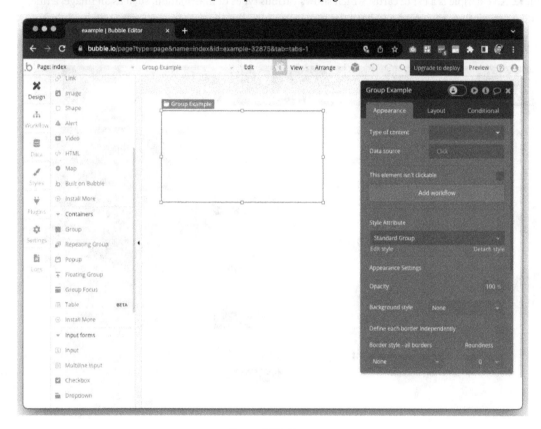

Figure 3.12: Group

If you know HTML, a Group component is the equivalent of a DIV component, meaning it will be used to create sections, big blocks of components on your page, and hold multiple little components together inside it. For instance, you can use a Group component to hold an image, a title, a description, and a CTA button, and call that entire group section; this is a very common example that is used in pretty much any landing page structure and would be a very specific and practical way to use Group components while building your pages. You can think of a group as an empty square or a box, which will store something inside it and will help you organize and structure your page layout. Group components are useful for creating big sections and blocks inside your pages, reusable elements, navigation bars, footers, and much more. Get familiar with Group components, because you are going to use them a lot!

RepeatingGroup

Sometimes, you are going to create components inside a page that will look similar and will have the same structure, the only difference being that the content in each component will be different and dynamic, meaning it will come from your database and change according to the records available there. An example is a list of cards, which shows various types of information, such as an image, a title, and a description; each card will have the same structure, but the data for each card will be different.

The **Bubble Editor** page with a `RepeatingGroup` component added to the page is shown here:

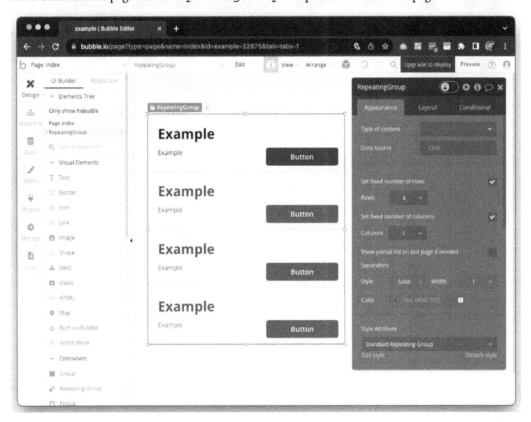

Figure 3.13: RepeatingGroup

`RepeatingGroup` is like the `Group` component but there is a key difference; this component is a dynamic element that enables you to display lists of data from your database on the components themselves; it will replicate the first `Group` component multiple times depending on the amount of data received from the database and how you configure it. To use it, you will only define and design the first component on the list, plug it into the database, and it will repeat itself to display as many items as you need on that list, bringing dynamic data from the database and creating copies of the same component, keeping the structure but changing the data. This is a very cool and powerful component that can be used to create dynamic lists and grids. You can customize the appearance and behavior of RepeatingGroup components to showcase various data in your app.

Popup

This element appears as a modal overlay on top of your page's content. The popup is used to display additional information, forms, or alerts without navigating to a new page. It can be useful for creating interactive and context-specific user experiences, such as login forms, notifications, or interactive menus.

The **Bubble Editor** page displaying a Popup element on the page is shown here:

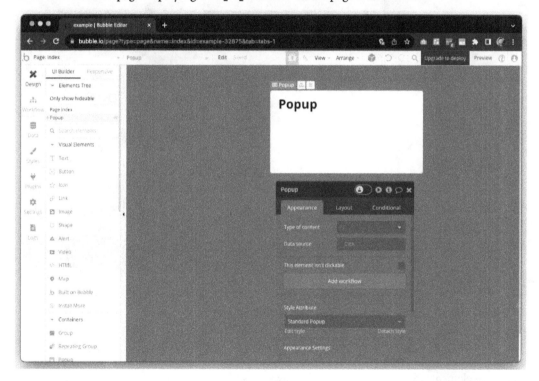

Figure 3.14: Popup

Note that when you add a popup to your page, it will first appear for you to edit, and then, once you click away, it will be automatically hidden. To reveal it, you will need to click the eye icon available in **Elements Tree**. Also, remember to set up a clickable action to review the popup, as it is, by default, hidden on the page load, so it won't just show up unless you create a workflow to make it visible.

FloatingGroup

Different from the Group container, this is a dynamic container that stays fixed in a specific position on your page as users scroll through your app. It is similar to the Group component in various aspects, apart from the fact this is going to stick to the page and stay there, on top of other existing components on your page.

The **Bubble Editor** page with a FloatingGroup component on the page is shown here:

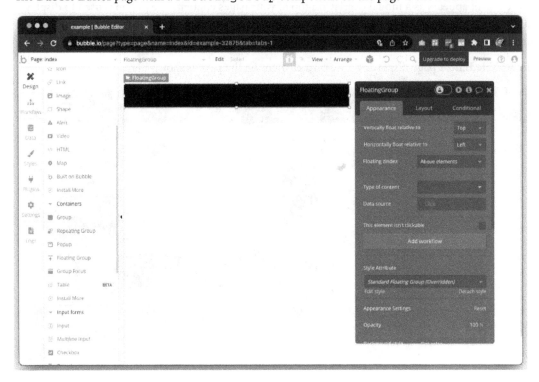

Figure 3.15: FloatingGroup

This feature is useful for creating sticky headers, navigation menus that are on top of other components, or persistent elements that remain visible and easily accessible to users, enhancing the navigation and user experience.

GroupFocus

This component works like a Group component but with a singular difference – GroupFocus is a hidden component by default, and it will only be visible when you specifically make it visible using a workflow action.

The **Bubble Editor** page with a `GroupFocus` component on the page is shown here:

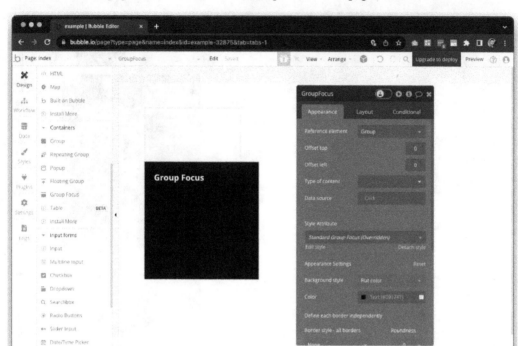

Figure 3.16: GroupFocus

It will also appear next to an existing component that will be used as a reference for its location. So, it can be useful for creating little windows and tabs that need to appear close to an existing component; for instance, if you want to create a dropdown menu, this can be a good option because it will be visible close to another existing visible component applied to the page. Another common usage of it is for displaying tooltips or providing context-specific information within your app.

Table

With this component, you can create structured and organized data and display it in your app. You can populate tables with data from your app's database or external sources, and customize the appearance, sorting, and filtering options.

The **Bubble Editor** page with a `Table` component added to the page is shown here:

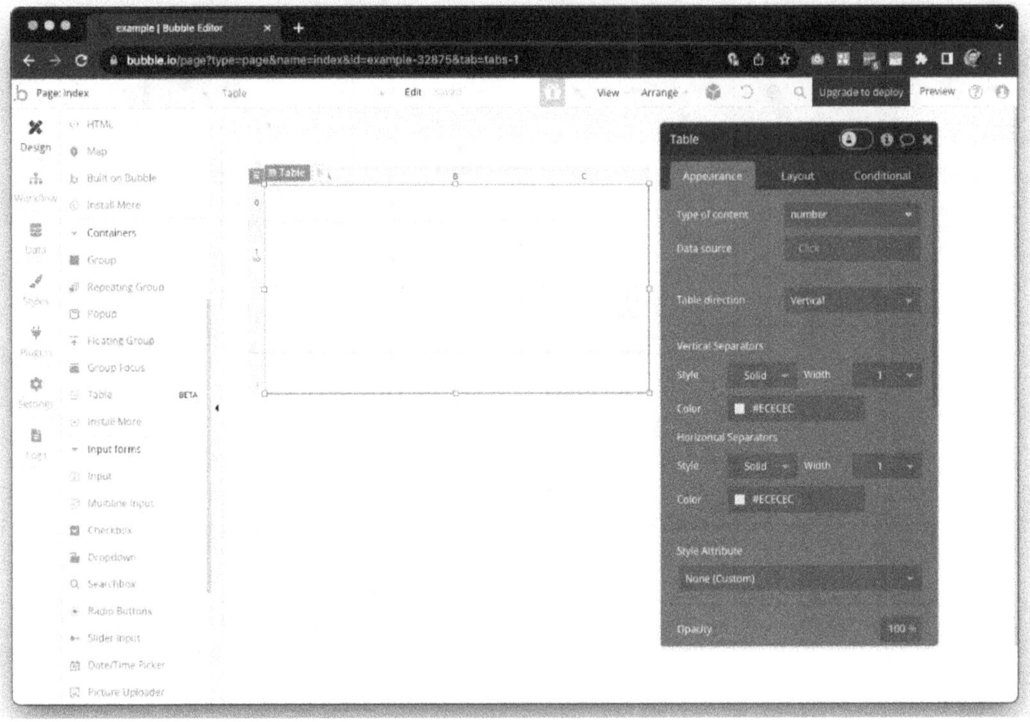

Figure 3.17: Table

Tables are excellent for presenting data in a tabular format, such as spreadsheets, leaderboards, or catalog listings. This is a newer component, and it brings more flexibility for lists with heavy loads of information. Some people would use `RepeatingGroup` components to display data lists like in a table, but sometimes the `Table` component can be better for that task. You can experience and see how they work and choose the best option for your type of project.

Input forms

The next category of components is the Input forms. Here, you are going to find fields that can be useful for creating forms and adding dynamic data to your databases. Without inputs, you wouldn't be able to create more complex types of applications inside Bubble, so make sure to learn how they work to be able to build user sign ups, logins, and much more cool stuff; let's dive in.

Input

When creating forms, the Input component is a fundamental element that allows users to add single-line text to your application, such as names and email addresses, and submit this data to your database.

The **Bubble Editor** page with an Input component on the page is shown here:

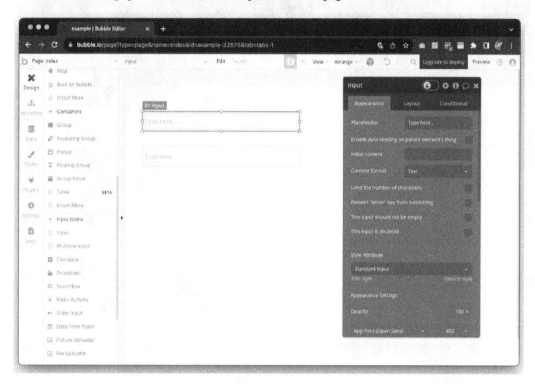

Figure 3.18: Input

Inside the editor, you can customize the Input fields, define placeholders, and specify data types for validation. The Input field is commonly used to create sign-up login forms, contact forms, searches and collect data from your users to perform actions and store data on your application's database.

Multiline input

This component extends the input capabilities by allowing users to enter longer text or multiline content.

The **Bubble Editor** page with a `MultilineInput` component on the page is shown here:

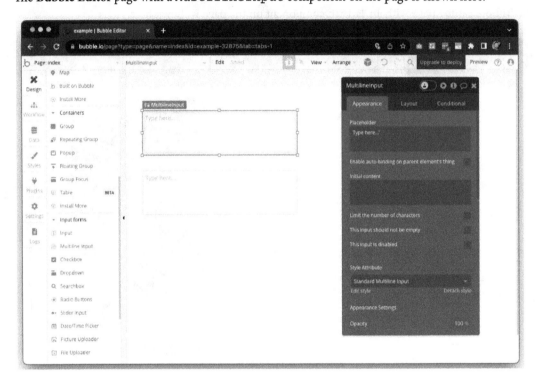

Figure 3.19: MultilineInput

The `MultilineInput` component is basically the same as the `Input` component, but with a bigger size and is used to collect more information at the same time. It is useful for capturing paragraphs, comments, descriptions, or any text that spans multiple lines.

Checkbox

Sometimes, while building forms, you need your users to make binary choices, where users can either select or deselect an option.

The **Bubble Editor** page with a `Checkbox` component on the page is shown here:

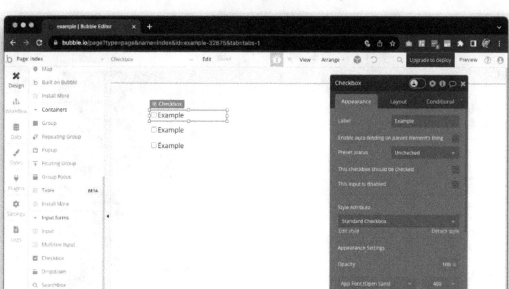

Figure 3.20: Checkbox

Checkboxes are commonly used for tasks such as accepting terms and conditions, toggling preferences, or selecting items from a list. Remember that checkboxes are used for questions that usually are answered with yes or no, and the answers can be stored on your database to remember what the user chose when filling out the form with checkboxes.

Dropdown

This component provides users with a list of selectable options in a drop-down menu format. This is usually used when you have a form, and the users can choose a single item from the list.

The **Bubble Editor** page with a `Dropdown` component is shown here:

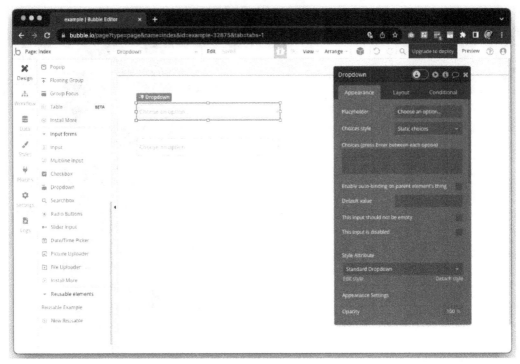

Figure 3.21: Dropdown

Usually, that list will come with predefined answers, so the users must select one, making dropdowns ideal for selecting from a predefined set of choices, such as categories, countries, or states. You can define the list of answers straight inside the component as a static list or pull the data from a database, making it a dynamic component.

SearchBox

Instead of using an Input component for searching, you can also use a search box, which is an input field combined with a search icon, designed for users to input search queries.

The **Bubble Editor** page with a `SearchBox` component on the page is shown here:

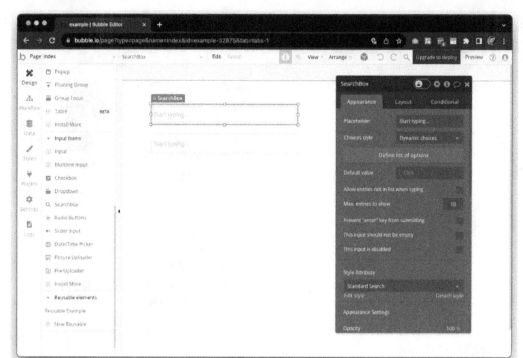

Figure 3.22: SearchBox

A search box enables users to search for specific content or filter data within your app, enhancing navigation and content discovery. You can combine the `SearchBox` component with other components, for instance, a `RepeatingGroup` component, to make it change the **List** items on that page based on what people are adding to the search box.

RadioButtons

Different from the checkbox, radio buttons are used to present a list of exclusive options where users can select only one choice at a time.

The **Bubble Editor** page with the `RadioButtons` components added to the page as shown here:

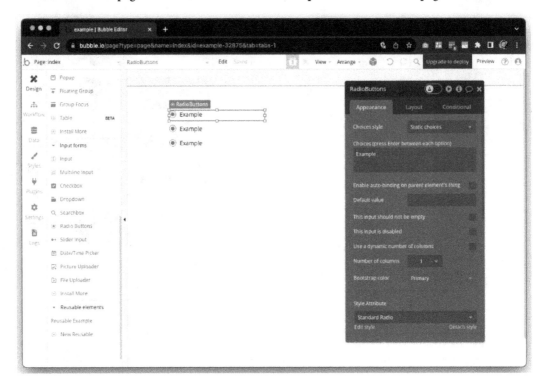

Figure 3.23: RadioButtons

They are suitable for scenarios where users need to make a single selection from a set of mutually exclusive options, such as gender, payment methods, or preferences. So, always remember that if the answer to a specific question needs to be unique and only one, use a RadioButtons component instead of a Checkbox component.

SliderInput

As the name suggests, this component allows users to select values from a predefined range by sliding a handle along a track.

The **Bubble Editor** page with a `SliderInput` component on the page is shown here:

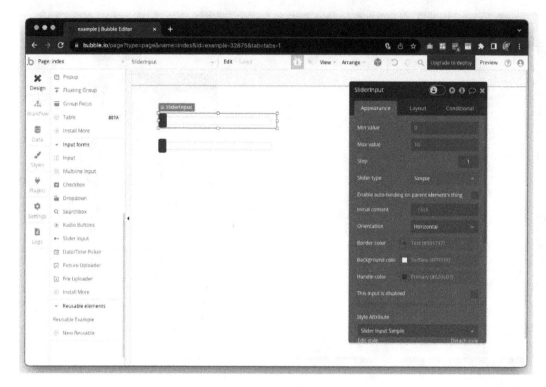

Figure 3.24: SliderInput

It is perfect for scenarios where users need to input numerical values within a specified range, such as price selection, volume adjustment, or rating input.

Date/TimePicker

If your forms require date or time selection, this component provides users with interactive calendars or clock interfaces for selecting dates and times.

The **Bubble Editor** page with a `Date/TimePicker` component on the page is shown here:

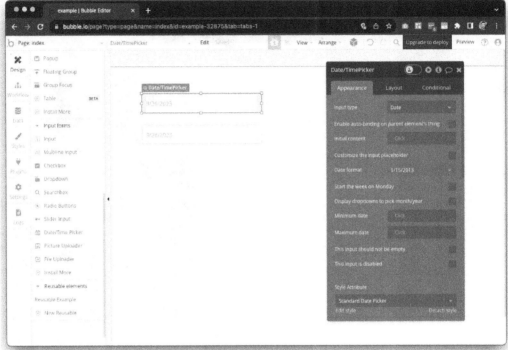

Figure 3.25: Date/Time Picker

They are essential for capturing date-related data, scheduling events, or setting deadlines within your app. This is a very useful component because it already comes with everything you need ready, so you just need to drag it to the page and configure how it will function.

PictureUploader

When you need the users of your application to add their photos or upload an image to your app, this component will help you do that.

The **Bubble Editor** page with a `PictureUploader` component on the page is shown here:

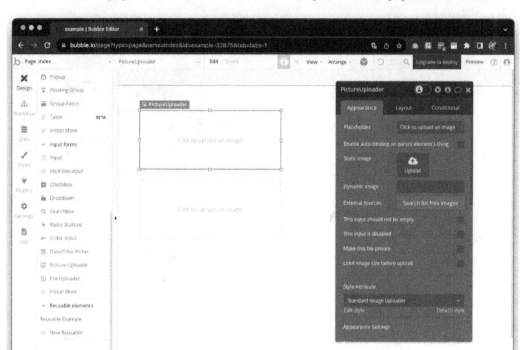

Figure 3.26: PictureUploader

With the `PictureUploader` component, users can select images from their devices or external sources and send them to your app's database. It is very customizable and easy to use; you can also control and specify image dimensions and compression, and make sure the files will be stored in an optimized way.

FileUploader

Similar to the `PictureUploader` component, this component allows users to upload various file types, not just images, including documents, PDFs, and spreadsheets. It's a versatile tool for handling file uploads and document management within your app.

The **Bubble Editor** page with a `FileUploader` component on the page is shown here:

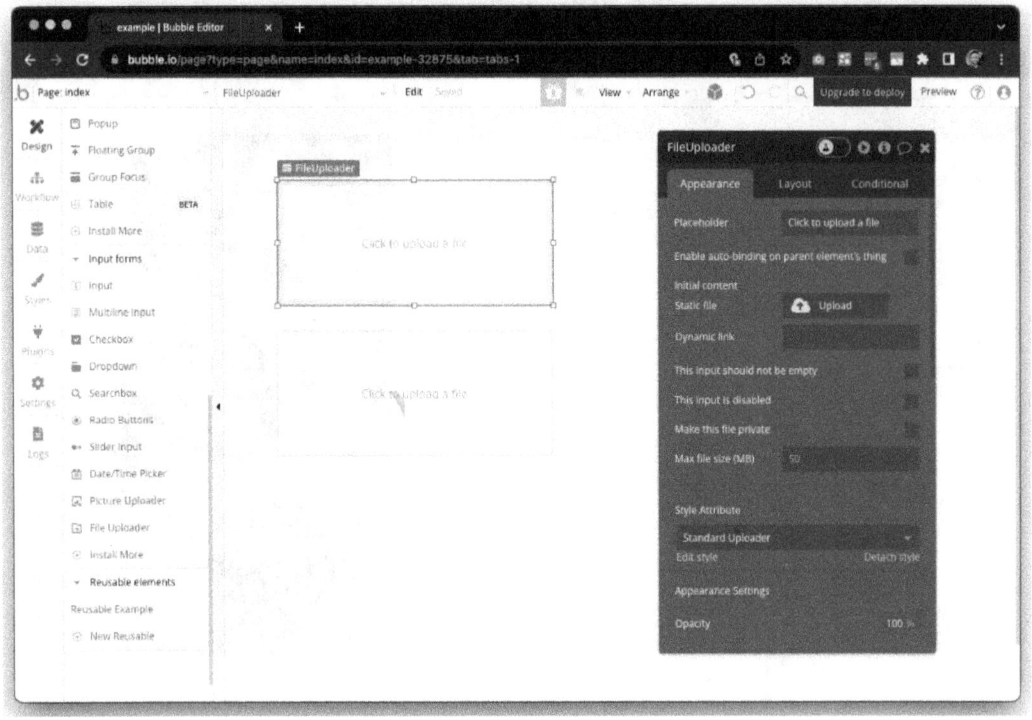

Figure 3.27: FileUploader

Now that you know all the components available inside the **UI Builder** tab, let's talk about reusable components and how to add custom components from the marketplace to extend Bubble's capabilities even further.

How to create a reusable element

A reusable element is a powerful feature in Bubble that allows you to create custom components or templates, which can be used across multiple pages of your app. Once created, these elements act as building blocks that you can easily insert into different parts of your app and can be easily maintainable. They are particularly useful for maintaining design consistency and functionality throughout your app, as any changes made to a reusable element are automatically reflected across all instances where it's used. This feature streamlines app development, encourages modular design, and saves time by eliminating the need to recreate similar elements on each page.

To create a new Reusable component, follow the outlined steps.

There are two ways of creating reusable components – the first is by clicking on **New Reusable**, and following these steps:

1. Click the **New Reusable** link.

Figure 3.28: New Reusable

2. Give the component a name or clone an existing one (duplicate).

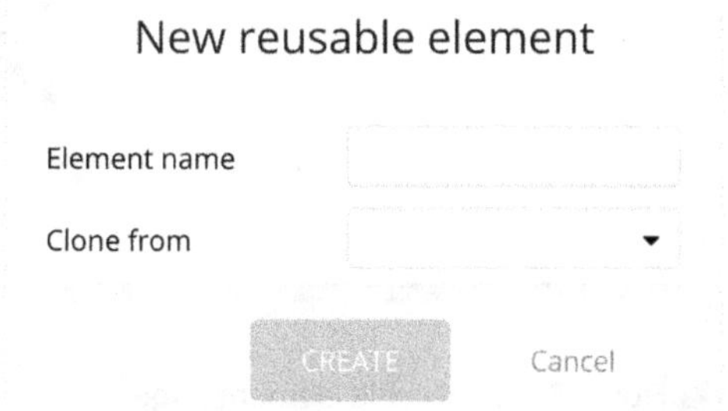

New reusable element

Element name	
Clone from	

CREATE   Cancel

Figure 3.29: Create a reusable element

3.   Once you click the **CREATE** button, you will be directed to the page where you need to design your customized reusable component.

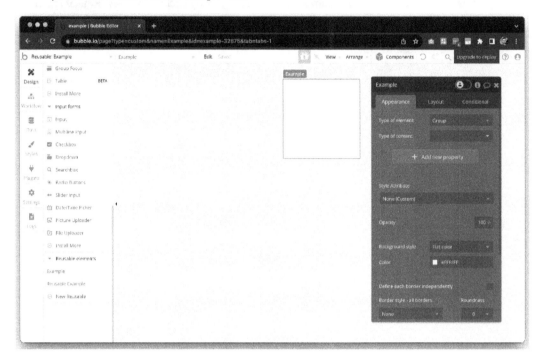

Figure 3.30: Reusable element page

This empty canvas represents your entire component – all you have to do is set up the page and add the UI elements you need inside this empty area.

After you are done, you can drag this newly created reusable element to the main pages you want it to be added to and you are good to go.

The second way of creating a reusable element is by first creating a component on any page and then turning it into a reusable component later. To do that, follow these steps:

1.   First, decide which component you will need to use across your entire application, for instance, a header, a footer, a single section, or any component.

2.   Using the components available inside the **UI Builder** tab, create the component you want to convert into a reusable component. You can do that on any page.

3.   Now, select the entire component, right-click, and choose the **Convert to a reusable element** option to create a new reusable component using this existing component. Like in step one, give the component a name.

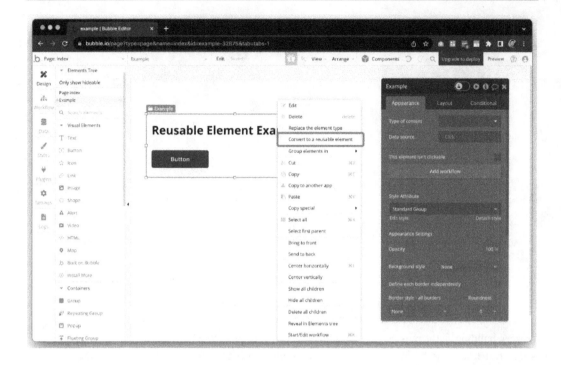

Figure 3.31: Creating a reusable element

4.   Now, you are going to be on the reusable components page. Your UI elements were automatically transported to the reusable element.

5.   Now, your reusable component is ready. You can go to the reusable elements section and see your new component available there; just drag it to the desired pages to apply it on the canvas.

Figure 3.32: The Reusable elements newly created components list

6.   Remember to delete the static version from your page, the one you created in *Step 1*.

7.   Now, every time you change it on the master page, it will also update everywhere else.

8. You can locate your reusable elements from the **Pages** tab in the top bar.

9. The top bar section example shows the **Pages** list and reusable elements available:

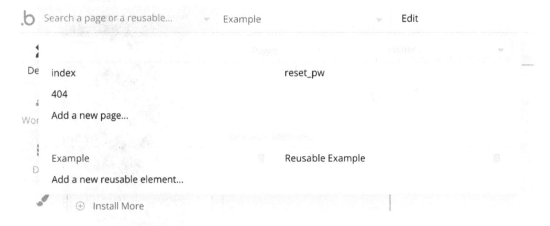

Figure 3.33: Top Bar, Pages, and Reusable Elements list

Now, keep building all the reusable elements you need on your project; with a set of reusable components, it will be much faster to build and maintain your application in the future if you are building it all by yourself or if you are working with a team.

Useful reusable elements – Header, Footer, Signup, and Login

When talking about reusable elements, if you are new to web development, I think it is important to know a few teams so you can get more familiar with the development world. If you are already an experienced professional, I am sure you can skip this part, but if you are not, I hope this can help.

Header

The Header component in Bubble serves as the top section of a web page or app interface. It typically contains navigation menus, branding elements, and key information that remains consistent across different pages or views of your app. Headers are essential for providing users with a clear and accessible way to navigate your app, ensuring a seamless and user-friendly experience.

Footer

The Footer component is located at the bottom of a web page or app interface. It often contains additional navigation links, copyright information, contact details, or other relevant content. Footers are valuable for providing users with access to important information, reinforcing branding, and enhancing overall usability. They help users find additional resources or get in touch with your app's creators.

Signup / Login Popup

The **Signup / Login Popup** is a UI element that appears as a modal overlay when users need to register or log in to your app. This component simplifies the authentication process by presenting users with dedicated forms for creating new accounts or entering existing credentials. Signup/login popups are common in many apps and websites, as they offer a convenient and secure way for users to access app features or protected content. They enhance user onboarding and data security by centralizing user authentication.

These are all examples of reusable elements you can have on your project to speed up your builds and maintain projects with ease.

How to install new components

By now, you already know that you can install new components on Bubble, meaning you can expand the possibilities of what you can build by using plugins developed by the Bubble community and Bubble developers out there.

Let's learn how to add new components to your Bubble Editor – it is very simple:

1. First, click the **Install More** button, as shown here:

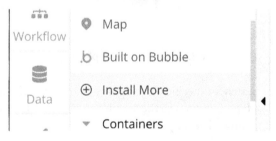

Figure 3.34: Install More

> **Note**
> There is no difference if you click the link under a specific category, it will open the same page (Plugins marketplace).

The Plugins Marketplace modal window is shown here:

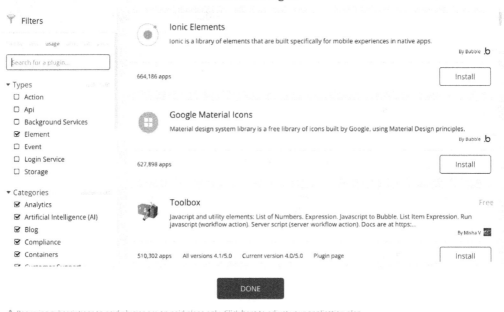

Figure 3.35: Plugins marketplace

2. From that list, you can search for specific UI component names; for instance, if you are looking for Icon components, type icons into the search bar. You can also use some filters to sort results by name, usage, rating, date, or price.

3. If you already know the specific name of a component or plugin you want to install, you can type it into the search bar or browse the list to find new available components to test. Bear in mind that some components are free to use and others are paid. Usually, there are two payment options: to pay a price to use it forever or subscribe to a monthly fee. Choose the option that makes more sense for you or just stick with the free one4

4. Now, once you have decided you want to install a new component, click **Install**. If it is a free component, it will automatically be installed, but if not, just continue to the checkout process and follow the step5

5. Once you've installed all the components you want, click **DONE** to close that page and go back to your application.

You will see the installed components available under the categories of your **UI Builder** sidebar.

Here is the UI components sidebar list with the newly installed components via plugins:

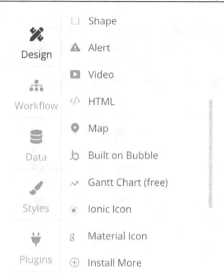

Figure 3.36: Newly installed components

If you need to manage your installed plugins and components, just head to the **Plugins** tab – you will see them all there. You can easily manage them, add more, or remove existing ones.

The **Plugins** tab shows all plugins installed:

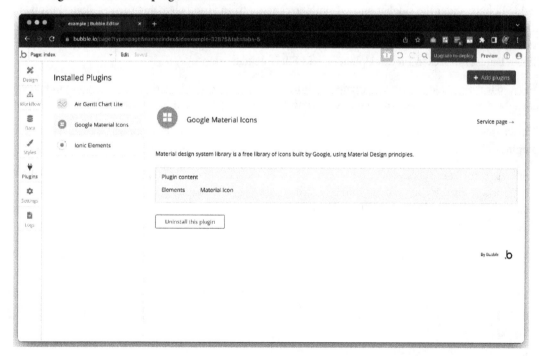

Figure 3.37: The Plugins management page

Plugins can help a lot while building any project inside Bubble because they can allow us to get extra features available inside Bubble that are not available natively, and luckily since we have millions of people out there using Bubble and building plugins, we have a lot of options to choose from and extend what is possible to build with this amazing tool. Now, a quick tip is to choose plugins carefully and not add too many to your project because it can make it run slow. So, make sure just to add plugins if you actually need them. Also, pay attention to who created the plugin, as some were created by the Bubble team, but some were created by other companies and individuals. When you install a new plugin, you rely on that plugin and who created it, so it is a dependency. Also, make sure you install plugins that are updated and not broken, and avoid installing plugins that can damage your app or cause security issues. Always check who is behind a plugin and the reviews of other Bubble users. If you've got yourself thinking, can I create my own plugin? Yes, of course, you can! This is a subject for another book, but know that it is possible to create and use your own plugins if needed.

Now that you know how to install new components using the plugins marketplace, take some time to explore the marketplace and also check the list of recommended plugins we prepared for you in future chapters. I will see you there soon!

Summary

In this chapter, we've covered a number of important concepts that will help you build a strong foundation while using Bubble. You learned about the Design tab and each category and UI component available inside Bubble under the sidebar and UI Builder, ranging from visual elements such as Text, Button, Icon, Link, and Image. You also learned about containers such as Group, RepeatingGroup, and Popup and form elements such as Input, MultilineInput, Checkbox, Dropdown, SearchBox, RadioButtons, Slider, Date/TimePicker, PictureUploader, and FileUploader.

You also learned what reusable elements are, how to create them, and how they work so you can take advantage of this useful feature to build faster and consistently maintain your applications.

This chapter also explained how to install new design components and how you can benefit from plugins to be able to extend the possibilities of what you can build with Bubble.

Understanding these concepts and properly knowing each element available in the Design tab will be important during your Bubble-building journey. In the next chapter, you are going to learn how to structure your ideas to start building your first Bubble app.

4

Building Your First Bubble App – The Planning Phase

In the previous chapters, you learned the foundation of Bubble components and the editor. Now, it is time to talk about planning. It might sound counterintuitive to not continue learning about the tool itself, but trust me – knowing how to use the tool is important, but knowing what to build is also very important. In this chapter, you are going to learn how to prepare and plan to build your no-code application so that when you're building it, you know exactly what to do. This approach will help you avoid getting lost inside the editor and building things that are not necessary initially, at the early stages of your project, reducing complexity and your chances of getting stuck.

In this chapter, we'll cover the following topics:

- Identifying and defining your target user
- Defining the problem you are going to solve (project goals)
- Outlining the desired functionalities (app structure)
- Mapping out user interactions and flows
- Sketching page structure (wireframe)
- Planning the data structure (database)
- Defining third-party services

Identifying and defining your target user

Before building anything inside Bubble, you should think, who am I building this for? A common mistake most people make is to build something first and then try to find someone to use it. Most of the time, it leads to building things no one wants. If not, it leads to building something that doesn't solve a problem – something that doesn't add real value to the users. That is a terrible thing. Imagine that you spent time building something and then discovered that it is useless or that no one wants to use it. This isn't nice, but it happens all the time. And trust me, people sometimes lose years building the wrong thing and also spend a lot of money during this process. So, avoid that at all costs.

How can you avoid that? Simply by thinking and planning your project before jumping into building things. The building part is fun; planning might not be, but it is important. If you want to succeed, planning is the first step and it will bring problems along with it.

No plan is perfect, but any plan is better than no plan at all.

By failing to prepare, you are preparing to fail.

– Benjamin Franklin

So, how do we start planning our application? If we want to use a user-centric approach while building a product (which I recommend), the first part of any plan is to think about the users. After all, you're building something to help someone, right? The user-centric approach means putting the user at the center – you are building a product, a solution, or a piece of software to help someone who has a problem that your software, solution, or app will potentially solve. And I mean potentially because you will only know if that works after testing.

So, start with the user in mind. Who is that user? Who would benefit from your idea or your product? Who are you building this for?

It might sound hard to define who that user is and pick one, but it is very important to do this. If you don't have clarity about who you are building this for, how can you know what to build exactly? If you just say or think you know it all because you are the user, there is a huge risk you will build the wrong thing, or you will build something just for yourself. If that is the case, that's fine, but if not, then reconsider. Remember that your goal is to provide value to your users and solve their problems effectively. If you don't know who they are, you won't be able to help them, and if you try to help everyone, you will end up helping no one. So, yeah – saying it is for everyone doesn't cut it.

Do you know who your user is?

Do you know enough about your user? If you don't know how to get started, there are a few options for defining who your users are or at least having a first draft to guide your project-building process so that you can start building and testing your product hypothesis.

There are two ways to do this. First, you can start with the user. Identify a user group or individual and research their problems, talk to them, and create a survey that will help you understand if they have a problem. Is that the same problem you are willing to solve, or a different one? During your conversations, does the problem appear naturally? Remember to never start conversations talking about your product or solution; engage with users in a broader conversation about themselves and their daily activities and routines and see if a problem arises from the conversation. If you discover that these types of users do have the same problem, then you're on the right path. If not, then these are probably not your users or your solution is not something they are looking for. At this point, you can choose to talk to other people or iterate on the idea, maybe even considering what you've learned from the conversations you've had. If this group of people share the same problem, you've found your user group. At this point, you need to understand their common characteristics and get to know them on a deeper level; this will help you build a better product. Make sure you save this information in a document, be it using a Word document or a spreadsheet. You can even use Airtable or any other software you like – just make sure to collect data about your users. To help with your user interviews, I've added a little structure that you can use as a reference.

The second approach is to start with the problem and then understand who the users with that problem are. Once you find a clear problem, you will be able to find who is willing to solve the problem, at which point you will know the user and the problem. When starting with the problem, you might find different groups of users with the same problem. Try to identify which group of users are more likely to use your product solution, what users are actively looking for a solution to their problem, and which users are more eager to buy anything that solves their problem. When researching users while starting with the problem, you can use this as a filter during interviews. If the person doesn't have that problem, you can potentially discard that interview and continue searching for users who have the problem. Another way to test the problem is to build a simple landing page or website to collect interested users. If you already know the problem and there is a real market need for a solution, you can use it to bring users to you instead of hunting for users anywhere. Once people start to register and show interest, you can pick the ones that are more engaged to talk to and learn more about them until you understand what types of users are more likely to use and buy your product. With this information, you have a clear target user to go after.

Another thing to understand during the interview process is if people are also using alternative solutions or solving that problem in other ways. Maybe they are solving the problem with another tool or built a hack to solve the problem. Whatever they are doing to solve the problem can give you insights into how to compete and overcome existing solutions with your product. You can also learn how happy users are with the current solutions and if they are willing to use something else instead.

No matter what method you choose, just know that it is very important to understand your users and their problems. With this information, you are more likely to build something of value that will cater to what they need. So, please, don't skip this step.

It is advisable to research topics such as user personas and problem statements so that you become familiar with these concepts at a more professional level. You can use tools such as Miro or Notion to document your findings and definitions. This way, you will have a place to go back to whenever you need to consult this information to make product decisions.

If you start with a problem or solution, you can use an idea you already have. Begin by identifying the problem your product or service aims to solve or the idea you think will solve that problem.

But there's a catch! If you start with an idea, there is a risk that the idea might not solve a real problem, so make sure the problem is real. What do I mean by real? Well, a problem that exists is a problem that everyone agrees on and states that they are facing. If that problem isn't mentioned, it is just a problem for you, so the solution or idea might be interesting but there won't be enough demand for it. If that is the case, then maybe you should look for another problem and focus on understanding the problem first and gathering ideas later. Another cool thing to research is the jobs-to-be-done methodology, which helps you understand what jobs people are trying to accomplish and what products they hire to get the job done.

If you understand the problem well enough and the idea you have solves the problem, you can start thinking about what pain point this solution will target. What need does it fulfill? Is it clear? Understanding your product's purpose is important while identifying potential users. To understand more about pain points and the value you are generating, I recommend taking a look at the value proposition canvas.

If you don't fully understand the user, the problem, and the idea, conduct thorough market research to gain insights about the users, the problems, the market, existing solutions and competitors, and potential user demographics. Look for gaps in the market that your product can fill.

If you serve more than one user and there is no problem, you can define a group of users as your target users – just make sure you group users with similar characteristics and goals. If you want to go one step further with your users' understanding process, you can use an empathy map to gather insights from your user interviews or even use personas.

Consider each group of users as one type of user as a way to guide your development decisions based on their needs, create features for each group, and decide how the product should work using the data you acquired during this process. This will change the way you build solutions.

User interview tips

Find users where they usually hang out. You can look at social networks, groups, and forums and even go offline and meet people where they are. If needed, try and schedule an interview – just show up and ask if they can spare any of their time.

Here is an example of an interview structure:

1. Start by understanding the user's context and establishing an empathetic connection.

2. Ask people to share their daily context, who they are, what they do, and where or how they typically engage in the activity you want to know more about. Connect with the individual's pain points. Be open to listening and paying attention.

3. Try to understand what jobs they are trying to accomplish. What task are they trying to accomplish? What was their objective in seeking that solution? What were they hoping to achieve? Did they encounter any challenges? When was the last time they searched for a solution to this problem?

4. Try to understand what they've already tried to address this problem. What tools have they used? Have they utilized any services or products, or have they found other ways to solve it? What steps have they taken to alleviate this pain? How painful is that problem to them? Ask

them if they could envision an ideal solution to that problem; what would it be? What does the ideal solution look like? (The *magic wand* question)

5.   Show gratitude and keep doors open for future engagement. Express gratitude for their time and keep the lines of communication open for potential future interactions. If they are your target user, you can ask them to schedule another interview, extend the time to dive deeper, and explore the possibility of them becoming paying customers. You can also ask them if they are open to a future contact to test your product when it is ready to use and build a list of interested and potential future customers. At the end, conclude the conversation with a positive attitude.

Make sure you take the time to evaluate the answers, organize discoveries, and use the data collected to make future interviews or to start shaping your product.

Defining the problem you are going to solve

Knowing who the user you are building for is important, but knowing the problem you are solving for this user is equally important – after all, people will use your software to get something done. And do you know what that is and how to make that happen? The whole reason why you are building this is to provide a better alternative – a better way to solve a problem – that people are solving differently and that you can help them make better, faster, and maybe cheaper with the correct software. Finding the problem involves a meticulous process of identifying a specific issue or challenge that a target audience faces and building a product/solution that helps them achieve their goal. This step is crucial because it lays the foundation for your product or service. If you build a product that doesn't help or solve the problem, people won't use it.

Defining the user and the problem not only guides your development efforts but also serves as a compass for your marketing and product decisions. The more precisely you can define the problem, the better equipped you are to build an effective solution that addresses your users' pain points and fulfills their needs. That is your goal!

A well-defined problem statement is the first step toward creating a product or service that truly resonates with your intended audience and has the potential to make a meaningful impact. That is why planning and having clarity around that is more important than building.

How can you find the problem? Start by conducting market research. If you know the users, talk to them, use surveys and interviews, and analyze existing data.

If you already know the problem, then talk to users to confirm that they also perceive this as a problem. Sometimes, we think we know the problem enough because sometimes, we have that problem ourselves, but it is important to understand if other people also have the same problem and if they are looking for a solution to that problem. If we decide to build software that only addresses our problem and the solution might not appeal to multiple users, then we are building an app just for ourselves. If that is the goal, then that's fine, but if your goal is to build a solution for multiple people, then make sure there is a substantial amount of people wanting to use it. Note that you need to talk to them and understand if they also share the same vision.

Listen to your potential users, empathize with their experiences, and identify recurring issues they face. Once you've gathered insights, distill them into a clear and concise problem statement that encapsulates the core issue your product aims to address. This problem statement will serve as your guiding light throughout the product development journey, ensuring that every feature and design choice aligns with solving the identified problem effectively.

What is a problem statement?

A problem statement is a clear and concise description of a specific issue or obstacle. It serves as a tool to concentrate the attention of you, your team, and other parties involved on the problem, its significance, and the individuals affected by it.

Here are a few examples of problem statements:

- **Facebook**:

  - **Customer**: Ivy League university students.

  - **Problem**: It is hard to find and connect with your friends online

- **Stripe**:

  - **Customer**: Developers at early-stage start-ups

  - **Problem**: Accepting payments online is stupidly complicated to set up

- **Airbnb**:

  - **Customer**: Conference attendees visiting San Francisco

  - **Problem**: Being broke and needing to pay higher rent

- **Instagram**:

  - **Customer**: Amateur photographers already on social media (source)

  - **Problem**: Mobile photos looked crappy; they were time-consuming to upload and difficult to share with friends

- **YouTube**:

  - **Customer**: People who made home videos

  - **Problem**: There was nowhere to share *home videos* on the internet

Some good advice is to not put a ton of pressure on yourself at first. Define the problem and keep going. If you identify that your definition isn't right, change it, iterate, make it better, and keep going – don't get stuck defining the perfect problem; instead, get a good understanding and start testing your hypothesis.

Regularly validate and refine this problem statement as you gather more data and feedback, ensuring that your product remains laser-focused on delivering a meaningful solution to your users. Keep going!

Outlining desired functionalities

Now that you have your users in mind and a clear problem statement, your goal is to think of a product that can turn your problem into a solution. How will you solve that problem? Make sure the features, actions, and activities can be performed while your software is used to connect with the needs your users have. In other words, make sure your app solves the problem at hand.

Sometimes, a problem you want to solve will be too big. A good idea here is to break that problem down into smaller pieces. Which part of that problem is the most painful? Don't try to tackle the whole thing at once; remember, you want to build fast and avoid complexity at first, especially if you are learning.

Write down the main features you will need to build, describe how this application would work, sketch a bit, and brainstorm. The idea is to find a way to solve the problem for that user using the software, using Bubble, using a no-code tool.

Make a list of all the features and prioritize them. You can list what's nice to have and a must-have feature, and even define what will be built first and what will be done in the future. Always prioritize what's more relevant and only build the most important features first; the rest can go in a roadmap. Remember that you want to build enough to test and validate and have small cycles of iteration to certify that you are building the right thing. If you discover you are failing, at least you can fix it and change your decisions without losing months or years building something that wouldn't work. The faster you discover and learn, the better.

Now, as an exercise, take a moment to list the main features and functionalities, then categorize them and pick just the top ones so that you can start building.

If you want to use something more structured to prioritize your features, I recommend using techniques such as RICE, ICE, KANO MODEL, EINSENHOWER MATRIX, and MOSCOW. Research more about it and choose the method you like the most, or simply use your own method. The most important thing is to plan, define, and move on.

Sketching your page structure

There are two things you need to know before building any project – what pages you are going to have, which is done via a sitemap, and what exactly you will display inside these pages, which is done via a wireframe or layout structure. Let's take a closer look:

1. Start by listing all the pages you think you are going to need. They can be for your website but also your software application, such as a dashboard or a page to edit user information (my profile), log in, sign up, and more.

2. Once you have a list of all the pages, you can start defining what each page will look like. There are various methods you can use to achieve that. One is to draw on a piece of paper and another would be using design tools and drawing a wireframe of that page.

    Here's an example of a wireframe:

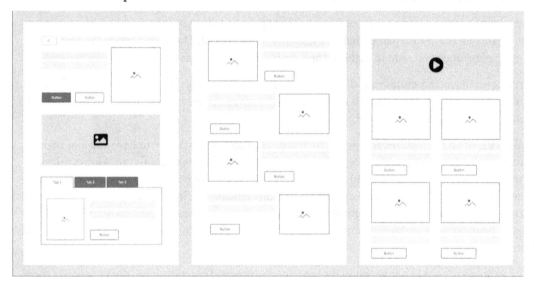

Figure 4.1: Wireframe example

You can also choose to not draw anything and just make a few notes about the page structure. If you are a designer, you can not only create a simple wireframe but a fully designed layout in Figma for instance, and then rebuild that pixel-perfect design inside Bubble after it is done. You can even follow your own process, but it is recommended that you at least follow these steps so that you can plan things before building as that is going to help you a lot.

Planning your data structure

If workflows are the brain of your application, then the database is the heart. Planning the database can make or break your application – you don't want to build the wrong database structure and discover that too late. It is OK to fail, but planning can help reduce errors. To plan your database, you can use specific tools created for that, such as LucidChart, or you can use pen and paper, Google spreadsheets, or tools such as Miro. It doesn't matter how complex or professional your database plan looks, so long as the structure makes sense. Now, databases are a complex area when using no-code tools since they come from the development world, and you are probably not going to be a database expert or become one overnight. Databases can become complex easily, so don't put too much pressure on yourself, especially at the beginning.

Before building your app, just try to write down what types of data you are going to store and how you can organize it logically. Usually, a database is composed of tables and columns; each table stores one piece of data and columns store information about that item on the table.

Here's a quick example:

Database

Users	Properties	Example
Email	Name	Item 1
Password	Photo	Item 2
Name	Description	Item 3
Avatar	Address	
Created Date	Size	
Properties	Beds	
	Baths	
	Owner	

Figure 4.2: Database planning

So, use that as a reference to build your own database structure. This is also called a database schema. You can search for more information about it on the web to get more familiar with what databases look like:

The following is an example of a simple database schema:

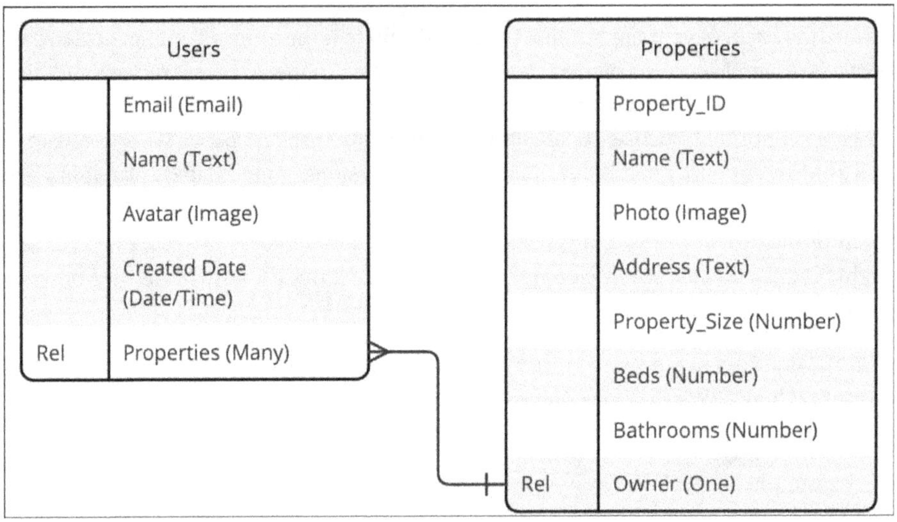

Figure 4.3: Simple database schema

At this point, you want to think about your app and sketch your database structure. It's OK if you need to change it later as this allows you to learn. If you think this is too difficult, maybe contact a professional to get some help – knowing how to properly structure your database is an essential part of building your application inside Bubble. If you are not going to build a dynamic website or just a simple static website, then you might not need a database, so you can skip this step when planning your application.

Defining third-party services

As mentioned in previous chapters, you can add plugins to your Bubble project to extend its capabilities and add extra resources to your application. Before starting your project, you can plan what plugins and extra resources you are thinking about using. This list can help you picture if you are going to use too many external resources and how much they will potentially cost. It will also give you an idea of how dependent your application will be on third-party services.

It is very good to be able to use plugins and external resources inside Bubble, and it is very nice that they allow us to use this feature. However, it is important to not add too many plugins to your project so that your application doesn't run slowly or even be less dependent on external resources that you can't control. I do recommend using plugins, but I also recommend not using too many if possible. That's why making a list of the external resources you are going to use can help you manage that before you jump into building. You can also continue updating this list during and after your project-building phase to keep track of what you are using.

The reason I am cautious about using plugins is that the more reliant on third-party services you are, the more expensive it can be to run your application and the more headaches you will have. For instance, if a plugin gets old or stops working and the developers who built it don't update or support it anymore, you might need to stop using it or consider an alternative. If that plugin is playing a huge part in your application, this situation might be more complex and costly to solve.

Plugins can help you build powerful applications, but make sure you use them wisely as they can cause issues and require management in the long run. Plugin makers might change the plugin and require you to do an update. This can also mean testing your application to ensure everything is still working properly. Bubble also makes regular changes to its system and releases new versions each year, which may cause plugins to break or become outdated, so pay attention to that. Don't worry – we are going to cover more about plugins in *Chapter 9*.

Another good practice can be to determine which plugins are more essential and core to your application so that you know how important it is and how difficult it would be to live without it or what impact it would have if it stopped working.

So, to summarize, do use plugins (they're awesome) – just manage them and keep track of things to make sure you are in control.

Summary

In this chapter, we explained the importance of planning and defining your project before building while considering your users and the problem you are trying to solve. This will help you avoid building the wrong thing and wasting your precious time.

We also explained how to map your app's features, define main user flows and pages, and create wireframes. Here, we provided tips on how to decide which plugins you are going to use to avoid making your application too slow.

Everything we've covered serves as an important guide to help you during the next steps, which will be primarily made inside the no-code tool.

Now that you know how to plan before building, in the next chapter, you will learn more about how Bubble works to create amazing designs, as well as how to work with styles and layouts. If you're a designer like me, you're going to like it!

5

Layouts and Styles

In the previous chapter, you learned how to plan and organize your app before building it. In this chapter, we are going to continue learning how to use Bubble, but this time, we will dive deeper into the layout and style options by exploring the **Styles** tab. You will learn the essentials to be able to change elements to your desired taste, choose colors and typography, and change all the little aspects of a component to be able to make your app look beautifully designed and aligned with your brand style and guidelines.

By adapting the design of your components and the layout structure, you can make your application look and feel more appealing to the users, creating an overall nicer user experience that can delight and engage visitors.

In this chapter, you will learn about the following topics:

- How to create layouts and main settings
- Layout customization options
- Design customization options
- Styling elements on the **Styles** tab

How to create layouts and main settings

Layouts are the foundation of any application. Your pages need a layout, even if it is a simple one, so it is very important to understand how layouts work inside Bubble, how to build one, and all the available settings that will allow you to build the perfect layout:

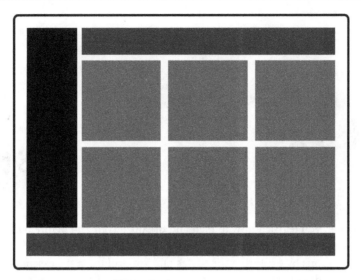

Figure 5.1: Layout structure

Layouts are built inside pages, so to build a layout, you need to start by adding container components to the page. As mentioned in previous chapters, containers are the components that will create the structure of your page and hold multiple components contained in a specific location. It's like building a house – the containers are the walls in that they hold everything together. Think of layouts as squared boxes that form a structure that will hold your components.

The most used component to build a page layout is the group element, which is available under the **Containers** category of your **UI Builder** sidebar. If you are familiar with HTML, a group is like a div – for example, a section or a container. To build a layout, you must add a couple of containers to the page and define how they should behave, how big or small a container should be, if it can grow or if it is fixed, how the components inside it should behave, and so on. You must also define what happens if there are multiple groups side by side, if they should stretch or divide the same space, how much internal space a component must have (we call this padding), how much external space it will have (we call this margins), if the component has a maximum or minimum size, and much more. This is what will define the page layout and how the page structure will behave. If you want to learn more about this, you can study topics such as responsive design, CSS grids, and flexboxes; these are the foundations that are used by developers when building a frontend. In the no-code space, these concepts are adapted and embedded inside tools such as Bubble. Visually, the idea is the same, but you will do it with no code.

These containers are usually invisible groups of components, but you can also add colors, a border, a background color, and so on. All these options are available inside the property editor under the **Appearance** tab. If you want to change how the layout works or how the component interacts with other components, you will find settings to change that under the **Layout** tab inside the property editor.

To start building your layout, it is important to configure your page settings first.

We can think of a page as a big group that comes added to our canvas by default. It is the higher group that will group everything else. So, it is important to configure the page before adding groups inside the page itself.

By clicking the page's name, you will see the property editor. From there, you will be able to change the page's settings, such as its appearance and layout.

The following screenshot shows the Bubble editor's **Design** tab, with the **Index** page selected:

Figure 5.2: Bubble editor – the Index page selected

In the property editor, you can click on the **Layout** tab and find various options. These options will help you define how your page layout will work and look.

The following figure shows the property editor and the different types of settings you can use to define your page layout:

Figure 5.3: The Index page selected – property editor options

Upon selecting a page, under the **Layout** tab, you will find different options to define how your page should behave. The first available option is the container layout type.

You can choose from **Fixed**, **Align to parent**, **Row**, and **Column**. Each of these options changes how the components you add to the page will be displayed.

Fixed

When using this option, the components will stay fixed at the place where you added them. This option allows you to just place components without thinking of a layout structure; the canvas is a free space and you can add components to a specific location and they will stay there. Even if you resize your browser window, they won't move or adapt:

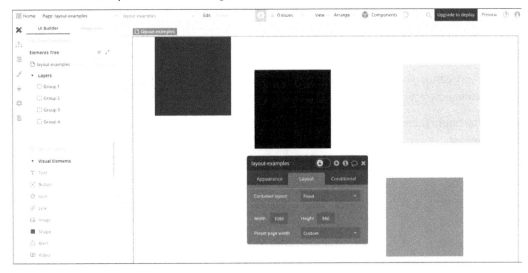

Figure 5.4: Index page with its container layout set to Fixed

Since this option is not responsive, it isn't the most used option, so be careful when using it. If you are going to use something fixed, make sure you know why you are using this option. If you build layouts using **Fixed** just because you think it is easier at the beginning, it can become a headache to make it responsive in the future.

Align to parent

When using this option, the components will behave according to the components that are their parents. For instance, when you add a button inside a group, the group is the parent component because it is wrapping the button; if there is no group, then the page itself is the parent. The parent is always the higher-level component that wraps other components. So, if you choose to align with the parent, it will consider the size of the parent as the new space to live inside; it will respect its boundary and adapt:

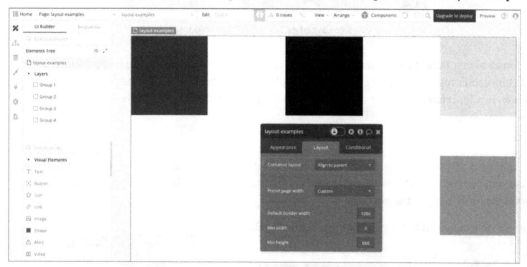

Figure 5.5: Index page with its container layout set to Align to parent

When choosing this option, if you click the component that was added inside the parent, such as a button or a group, you will see a few options under the property editor (the **Layout** tab) so that you can choose where you want that component to float to. Also, notice how this option eliminates any internal margin or padding between the component and the parent. If you drag and drop components, you will notice how they can move around specific areas and respect the space of other existing components. Just be careful – if you add two components to the same side, they will overlap each other.

The following screenshot depicts the property editor showing the **Layout** tab and its **Parent container type** options:

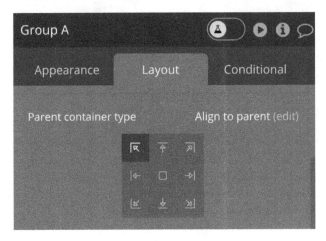

Figure 5.6: Layout options for Align to parent

Upon choosing **Parent container type**, you can choose if your component will be positioned at the left top, center top, right top, right middle, right bottom, center bottom, left bottom, left center, or center. Just for reference, this option is similar to the position absolute in HTML/CSS.

Row

When using this option, your components will behave as if there is an invisible row guiding them. They will be stacked horizontally, side by side, until there is no space left on the page width, at which point they will break to the next row.

Here is an example of index page settings defined as **Row**:

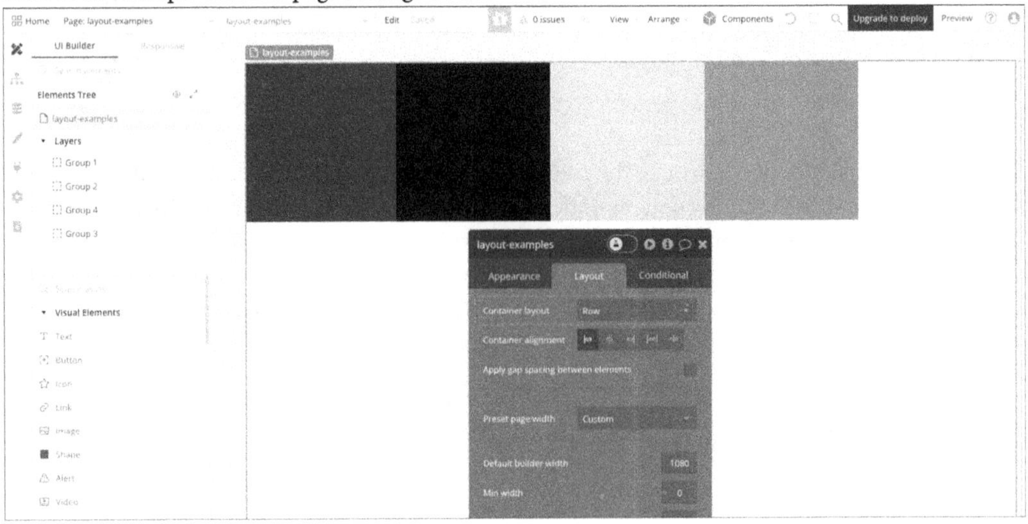

![Index page with container layout set to Row showing the UI Builder interface with Elements Tree, layout-examples, and a layout settings panel]

Figure 5.7: Index page with its container layout set to Row

As you can see, all the components are organized by row. Once you select the **Row** option, there are a few extra options available for you to choose under **Container Alignment**. These new options will allow you to define if the row is aligned to the left, right, or center, or even make the component spread throughout the page width, adding gaps between them. This can be used to build fluid layouts that balance components inside the page area as it grows or shrinks. I encourage you to build a similar page and play with these options to get familiar with them.

Column

When using this option, the components will behave as if there is an invisible column guiding them. They will be stacked vertically, one on top of each other, pushing the page height to make them fit.

Here's an example of index page settings defined as **Column**:

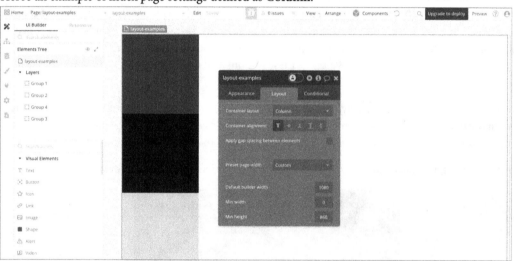

Figure 5.8: Index page with its container layout set to Column

The row and column options are very similar. The main difference is the orientation – the row will be horizontal and the column will be vertical. This option is the most used for building website layouts. Since your page will contain a lot of blocks, it makes a lot of sense to use columns and have multiple big blocks stacked on top of each other, serving as big containers for each section of your page.

Make sure that you test and play with this option. Inside each component that's added to the page, once you click on one and check the property editor, you will see that there are some new options to choose from.

The following screenshot shows these options:

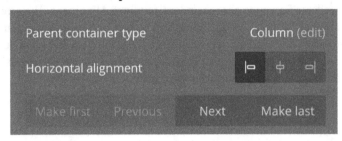

Figure 5.9: Alignment options for Column

These options can help you define how a particular component will behave inside its parent. You can choose if it will be aligned to the left, center, or right. If you want to change the order of a specific component, you can click buttons to make it first, switch with the previous or next, or make it last. These options can be helpful if you want to switch an entire section from a page with another one at the top. I encourage you to give it a try as these options will be very helpful when you're building layouts in Bubble.

With that, we've covered various layout options and settings. However, there is another option that is very important for you to know about: **Preset page width**. Once you click a page's name and go to its **Properties** panel, under the **Layout** tab, you will see a place to choose a preset page width.

The following figure shows the options that are available under **Preset page width**:

Figure 5.10: Preset page width

Here, you can choose from **Custom**, **Full width**, **Centered**, or **Mobile**. By default, it is set to **Custom**. This means you can define a custom size for your project pages. As you can see, you can define **Width**, which is set to 1,080 pixels, and **Height**, which is set to 767 pixels. Note that **Container layout** is set to **Fixed**, but as we've seen before, you can choose other options, such as **Row** and **Container**. If you select one of these options, things will change. Below **Preset page width**, you will see a couple of extra fields: **Default builder width**, **Min width**, and **Min height**. These values represent your page, the white central canvas inside the editor, this is the main area you will visualize and use when building inside Bubble.

Because you've selected different page settings, you will have different options that correspond to how that page setting functions. Don't worry too much about it – just focus on defining the page container layout and the page width. Usually, when building with Bubble, you are going to use the **Column** style most of the time. The page width can be 1,080, but this is not a fixed value; you can choose to build your pages a bit bigger or smaller. If you are a designer, set this value based on the grid you decided to use. Grids are always based on the number of columns you want to use – for instance, 6, 8, or 12 columns – and the gutter size, meaning the space between the columns. For instance, it can be a 12-column grid that's 960 pixels, a bit bigger at 1,170px or 1,280 pixels, or if you prefer a larger grid that's set to 1440 pixels or even more. It also depends if you're building for the web, tablet, or mobile, and if your website will be responsive or not. Sometimes your website doesn't need to be responsive at first or just target one specific type of device. You can also decide which approach you prefer, for instance, choose if your project will be mobile first or web first, meaning what size you will prioritize. If you choose mobile first, you will decide to start with the smaller viewport and grid size and adapt as it grows on larger devices, if you decide to make it web first, then it is the opposite idea, it will start adapted for web and desktop sizes and adapt as devices shrink. If you're not sure, talk to a professional web designer or developer or research the web to find out about the best practices before building your layout. This is very important. The height is usually flexible as your page will grow based on the amount of components and sections you add to it, so just add more as the page grows. For the height, just go with the flow and don't worry too much about it. **Min width** is only required if you want to define that your page will not be smaller than a certain size. You can keep it set to zero, but if you want to add something, 320 pixels is a good number.

By changing these options, you can define how your page will behave, which, in turn, means you can define how the components inside the page will behave and react as well. So, it is very important to define the page so that you can create cool layouts using groups later on. Once you start adding groups to your page, if you click on the group component, you will see similar options in the property editor panel (under the **Layout** tab). You can use the options that are available to determine how the group will behave inside your page so that you can start structuring your layout using custom settings. Here's an example:

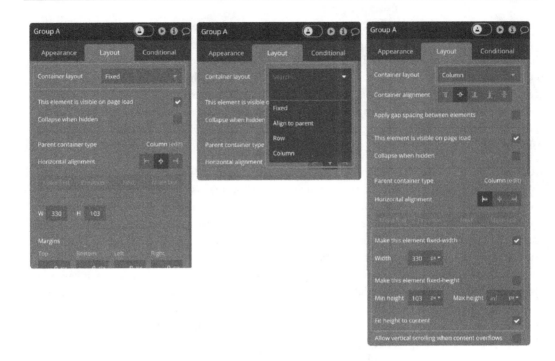

Figure 5.11: Group settings property editor

By default, once you add a new group to the page, it will be defined as **Fixed**. However, as mentioned previously, you can also choose other options, such as **Align to parent**, **Row**, and **Column**. If you choose one of these options, note how new settings will appear to let you customize the group component. You can define the alignment, apply gap spacing, collapse when hidden, select the horizontal alignment, make the element fluid or fixed width, and define width and height values. When defining the width, you can choose a pixel percentage, so if you want to ensure your container grows with the page, you can add a value of 100%, for instance. These are all the options you need to master to create awesome layouts!

When working on a fixed element, you will be able to define a specific size for the element by defining the width (W), height (H), position from the left (X), and position from the top (Y). If you select other options for these values, they will change accordingly. Play with them to learn about their differences.

You can also define margins and paddings. The margin is the space between external components and can be used to create a gap between two groups next to each other, for example. Padding is the internal space of an element, and it will determine how other components inside it behave and if they have a space between the edge or not. You can define the top, bottom, left, and right margin or padding independently if needed by adding a pixel value of your choice.

The following screenshot shows the **Margins** and **Padding** settings under the property editor:

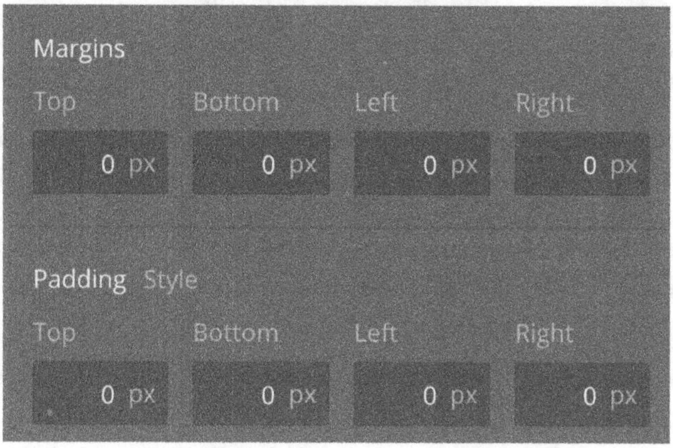

Figure 5.12: The property editor's Margins and Padding settings

Margins and paddings are very helpful when building layouts. Remember to add property spacing between elements as this is what makes your design breathe and be more professional. Don't be afraid to add white space as it is very important, especially if you are looking to build a clean, minimalist, or modern layout style.

Please keep in mind that the parent element also plays a huge role in the component's behavior, as well as how the settings that are available inside the property editor's layout tab are used. The parent is the component that holds the group – for instance, the page itself – or another group in case this group is inside another group. The parent's definitions also impact how the element behaves and what settings are visible inside the property editor. So, sometimes, you will have to check the parent first, configure it, and then go back to the child component – for instance, the group. This might sound complicated, but once you play with it, it will make sense.

Now that you understand how the components can be configured so that you can build a page layout, let's talk about appearance settings and styles so that you can make your designs not only structured but look good.

Layout customization options

Besides configuring how a component will behave in terms of layout structure, you can also define how it will look. This is important because we want to create functional applications that have a nice visual that appeals to our users.

To style any component inside Bubble, you can click on an existing element on the page and locate the **Appearance** tab of the property editor. Each component allows you to change slightly different settings, but most of them have pretty much the same options, such as background style, opacity, border style, and shadow style. Some other components allow you to select fonts, colors, and more.

Since we are talking about containers, it is common that we are more interested in defining the design of backgrounds, borders, and more. Once we click on a component, we can easily find all these available options. We can define these styles independently for each container component on the page or we can use a predefined style, which is a global style that can be shared across various components inside your Bubble application. Just locate the **Style Attribute** area and select one of the options available. If no option matches your desired goal, you can create a new one from scratch.

Using styles can be very helpful because you will be able to ensure consistency across your application. If you ever change the style, it will change on all the components that share that same style, saving you a bunch of time when you're updating your designs.

To manage existing styles, you can head over to the **Styles** tab. From here, it is possible to edit existing ones and delete and create new ones. You will see how the **Styles** tab works in more detail in the next few sections of this chapter.

Styling elements on the Styles tab

Bubble has a very cool area dedicated to managing the settings of your UI components under the **Styles** tab. It is very useful because you can organize and define how your components will look like and have styles applied to all the components that are using the same style across your entire application. If you are a designer, think of it as a design system. Having a place where you can centralize your component styles is great because you will be able to maintain consistency across your entire application, meaning the components will look the same and be standardized, preventing the user from getting confused about what a component does and how it works while also making sure your brand guidelines are followed accordingly. It is also very easy to maintain since you can change things once and these changes will be replicated throughout your entire application where components are using the same style.

There are two main tabs under the **Styles** tab. The first one is **Element style**, which contains a list of all the existing component styles you have available for your Bubble project.

The following figure shows **Element styles** under the **Styles** tab:

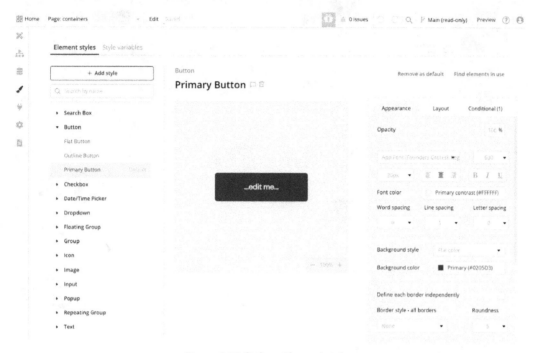

Figure 5.13: Styles – Element styles

The styles list is flexible and comes with some pre-defined styles by default, but you are free to add or remove them as you wish. To add a new style, simply click **Add style**. Give the style a name and choose which category it should belong to.

Notice that close to the **Element styles** tab at the top, there is a second tab called **Style variables**. Variables are like design tokens, if you are familiar with this term; if not, just think of variables as little design definitions that can go inside the styles themselves, such as a font family and colors. They are attributes that can be associated with a specific component style. However, once you change them, they change globally in every style that they are applied to, something that's very helpful. Inside the **Style variables** tab, you can create various font variables and color variables. Later, you can pick these options and group various settings under a specific style. We are going to learn more about this later. For now, just get familiar with this page.

Here's the **Styles** tab's the **Style variables** section:

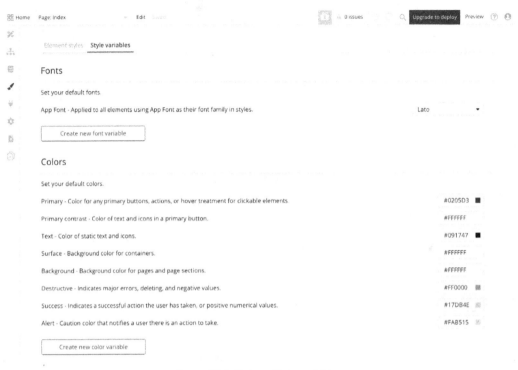

Figure 5.14: Styles – Style variables

To summarize, variables are the little options, like atoms, and styles are a group of variables and other design definitions. Together, they define what a complete component will look like; think of a style as a molecule. If you want to learn more about this, research more about design systems and the atomic design concept by Brad Frost.

Element styles

Now, let's dive deeper into the **Element styles** tab so that you can learn how it works and where to find every setting to make your applications look amazing.

Here's the **Element styles** tab:

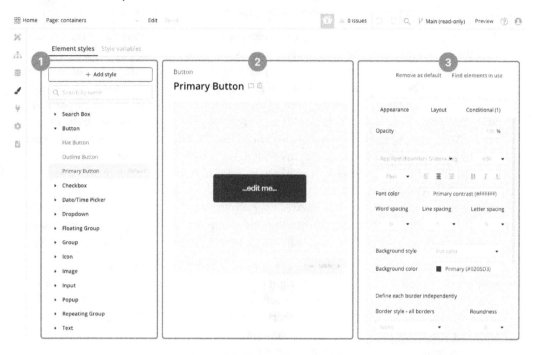

Figure 5.15: Styles – Element styles

Here's a quick breakdown of **Element styles** and each main area you should know about:

- On the left-hand side, at the top of the list, you will see a button to add new styles, a search filter, and a list of categories with an arrow next to it. The button to add new styles is straightforward – just click on it, give the style a name, and choose a category; your new style will be created! The search box lets you type in keywords so that you can find specific types of components. As we saw in previous chapters, we have all sorts of components in our application, such as alerts, buttons, text, inputs, groups, and more. If you wish to visualize the list of styles but just for buttons or a specific style name you've created, for instance, you can use the search box as a filter and just type in the keyword – it will show the available styles that match your search. Another option to locate existing styles is to click the category name on the styles list, below the search bar, and open all the styles that are available under a category such as **Button**. Once you click a category name, it will expose all the existing styles that belong to that specific category.

  When you create a new project, by default, your Bubble project will already come with pre-made style categories that group common components such as buttons, checkboxes, inputs, images, groups, and more. You can remove or add new styles as you wish. If you are an organized person and name your styles correctly, they will be very easy to find. If not, I recommend naming your styles comprehensively and choosing the proper category they fit in. I'm sure this will help you in the future when you need to do some maintenance work or change styles.

- At the center of the page, you will receive a preview of the selected component. This represents what that component looks like with the current styles applied to it. Here, you can see the category it belongs to and its name at the top. Next to the name, there are two little icons. The first icon looks like a text bubble. Upon clicking it, you get the option to add a note to that component style. This can be useful when you have a team and need to share information about a particular style. The second icon is next to the style's name and looks like a trash bin. It can be used to delete that style completely. Be careful – after the style is deleted, it can't be recovered. When previewing the style, you can also zoom in or out to see it in more detail; you can also click and interact with some components to visualize how they will behave in different scenarios – for instance, a button when it is hovered or clicked, a dropdown when it's opened, an input when it's focused, and more. You're encouraged to not only preview the components under the **Styles** tab but to also apply them to actual pages and run the application using preview mode to see how they are working and look, especially next to other components and in a page context.

- There's another feature you should know about that is related to styles. When a component is selected, let's say a button, at the top right of the screen, you will find two links: **Remove as default** and **Find elements in use**. Sometimes, rather than **Remove as default**, the first link will be **Set as default**. How does it work? If the selected style item's name has a default indicator at the side (look at the left sidebar, under the available categories), this means that this is the default style for that kind of component - let's say it's a button. Every time you add a new button to the page, it will have that default style. If the component you are editing is not the default, you can make it the default by clicking **Set as default**. So, depending on the status of that component, you can either set it as the default or remove it as the default. You get the idea. **Find elements in use** acts as a search on steroids that you can use to locate areas and components of your application that are using that specific style. This is useful because you can control how your style impacts changes globally and know what elements and pages will be affected in case you decide to change a particular style.

- At the very right-hand side of the page, you will see the **Appearance** tab. This window separates settings into three different tabs – one for appearance, one for layout, and another for conditionals. All of these are related to the selected components and style. Let's learn more about them.

The Appearance tab

First, we have the **Appearance** tab, which allows you to customize all the styles that are available for that kind of component. It will change accordingly, so if it is a button, it is going to show a few options, if it is text, it will show other options, if it is a map, it will show different options, and so on.

This is the most used tab as it allows you to define how your style will look.

Here's an example of the **Appearance** tab under the **Styles** tab:

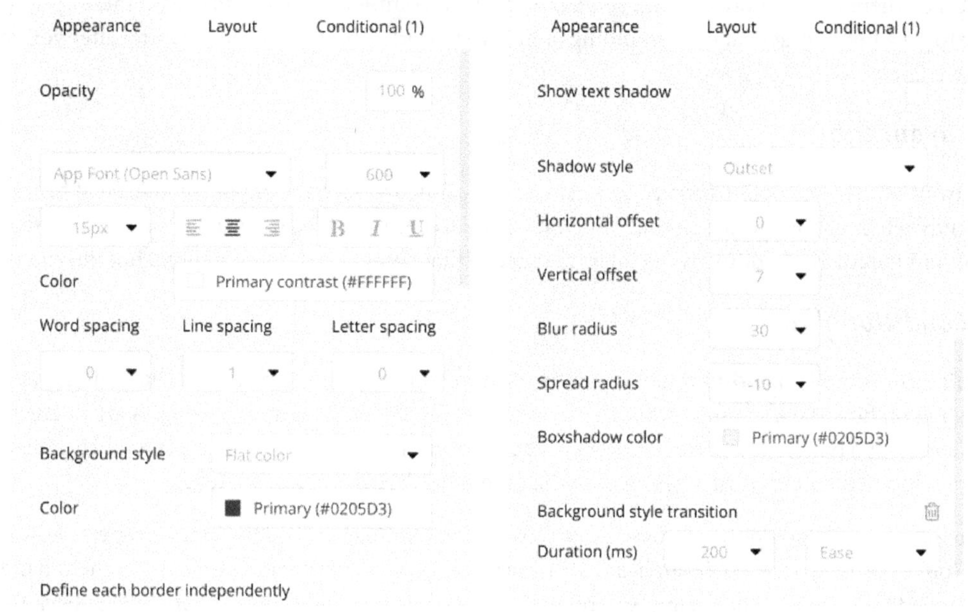

Appearance	Layout	Conditional (1)
Opacity		100 %

App Font (Open Sans) ▼   600 ▼

15px ▼   ≡ ≡ ≡   B *I* U̲

Color   Primary contrast (#FFFFFF)

Word spacing	Line spacing	Letter spacing
0 ▼	1 ▼	0 ▼

Background style   Flat color ▼

Color   ■ Primary (#0205D3)

Define each border independently

Appearance	Layout	Conditional (1)
Show text shadow		
Shadow style	Outset	▼
Horizontal offset	0	▼
Vertical offset	7	▼
Blur radius	30	▼
Spread radius	-10	▼
Boxshadow color	Primary (#0205D3)	
Background style transition		🗑
Duration (ms)	200 ▼	Ease ▼

Figure 5.16: The Appearance tab's settings

The first option in this tab is **Opacity**. This defines how transparent your component is. Remember that this is for the whole component, not just the background. If you want just the background to be transparent, you can pick this option when you're selecting a background color.

You also have options to define the font family. Here, you can pick a font variable, the font's weight and size, the arrangement (left, centered, or right), or if it is bold, italic, or underlined. You can also pick the font color, which can also be a variable. There are also options to define word spacing, line spacing, and letter spacing.

There's also the **Background style** option. Here, you can choose from **Flat color**, **Gradient**, and **Image**. You can also choose color variables here. At that, you can define borders, separate border styles, or configure all four corners at once.

Finally, you can configure shadow styles and add shadows for text.

As mentioned previously, the options will change according to the type of element being used. Sometimes, there will be more options, sometimes less. I recommend that you play with these options a little to get a better practical understanding of how they work. This will become natural after you've created a few styles.

The Layout tab

The **Layout** tab displays options related to what you can configure on that component to make it behave differently when applied to the layout. When it comes to components, it's usually the padding that can be configured. The padding is the internal space of that component, like a margin, but internally.

The Conditional tab

The third tab is the **Conditional** tab. It can be used to configure custom actions that will happen depending on how the users interact with a component. Some very common examples are to create conditionals for a button and to add a hover effect so that, for instance, when a user places their mouse cursor on top of a button, it changes color, indicating it has been selected or clicked.

As you may have already noticed, these three tabs are very similar to the tabs in the property editor when you're editing your Bubble application inside the main design canvas. Once a component has a style applied to it, most of the design features are no longer visible under the property editor because they are inherited from the global style. So, remember, every time you use a global style, always configure the component settings inside the **Styles** tab, not inside the component itself under the property editor that shows during the layout building phase when working on the design tab.

Shortcuts

Bubble has a few shortcuts you can use to speed up your development process. Memorizing and getting familiar with a few shortcuts can help you during your layout building process, that is why it is recommended to learn and use a few. To find shortcuts head to your Bubble editor, at the top bar, right next to the preview button there is a magnifying glass icon, click it and a new window will open, there you can find a link to review the shortcuts list.

The bubble shortcuts list as shown below:

Shortcuts List

Preview your app, equivalent to clicking on PREVIEW	**Ctrl+P**
Switch between the Design, Workflow and Data tabs	**Ctrl+T**
Select the element under the current element	**CMD + Click**
Resize the current element symmetrically	**CMD + Drag**
Resize and keep proportions constant	**Shift + Drag**
Copy current element, action or event	**Ctrl+C**
Paste current element, action or event	**Ctrl+V**
Cut current element, action or event	**Ctrl+X**
Copy current element's formatting	**Ctrl+Shift+C**
Paste formatting to current element	**Ctrl+Shift+V**
Duplicate current element	**Ctrl+D**
Show/Edit workflow for the current element	**Ctrl+K**
Select all elements on the page	**Ctrl+A**
Group the selected elements and move them in a new group	**Ctrl+G**
Center current element relatively to the parent	**Ctrl+E**
Turn this text element bold	**Ctrl+B**
Turn this text element italic	**Ctrl+I**
Turn this text element underlined	**Ctrl+U**
Open the PE	**Ctrl+**

Figure 5.17: Bubble shortcuts list

One of the shortcuts I would like to highlight is the Ctrl+P that can be used to preview your application, it is very handy, instead of clicking the button you can simply use your keyboard. Another one that can help is the Ctrl+E shortcut, that can be used to center element on a page or inside a group. There are just a few shortcuts available, but some can be very helpful to learn, I recommend you spend some time playing with them until they become second nature. You will see a great improvement in your layout building performance, it will pay off. If you are on a MAC, instead of Ctrl, hit Command. To copy one existing element style to another you can use the shortcut Ctrl + Shift + C and to copy the style of the selected component, then select the new one that will receive the style and hit Ctrl + Shift + V to paste it on the new component. It can be very useful! Now let's continue learning about styles and variables.

Style variables

Now, let's dive into the **Style variables** tab. The first section is **Fonts**. There, you can define how many fonts you want to add. Later, when creating styles on the first tab (**Element styles**), you will be able to select these fonts. So, if you change them later, they will be automatically synced with the styles that are using it. This is a useful feature. For fonts, Bubble users have the Google Fonts library, which is free and open source. When adding a new font variable, you can pick one from the library and choose the font name you wish to use. If you don't know about Google Fonts, just google it and get familiar with all the font options that are available so that when you add a new font variable, you already know the name of the font you want to use.

If you want to use a custom font, you can go to **Settings** > **General** > **Custom fonts**. It is very easy to set up, and you can find more information about how to do it by clicking the link to go to the Bubble documentation, which explains how to configure a custom font.

The second part is the **Colors** section. It works similarly in that you can create colors and choose a name for each color. Later, when creating styles, you can pick these colors from the color swatch. If they change in the future, they automatically update everywhere.

To create a new font variable or color variable, simply click the button **Create new font variable**, define the variable font or desired color, and give it a name. This is very simple and easy to do.

Adding styles to the components on the Design tab

After creating styles, you need to start using them. This part is very simple: go to the **Design** tab, add a component to the page, and click on it.

Once the property editor is selected, it will show an option under the **Appearances** tab called **Style**. Pick the desired style from that list and it will be applied to your component. Remember to make use of the default style option and assign a default style to your most used components, such as buttons, texts, and so on. This can help you speed up your building process because once you add a component to the page, it will have the default style applied to it:

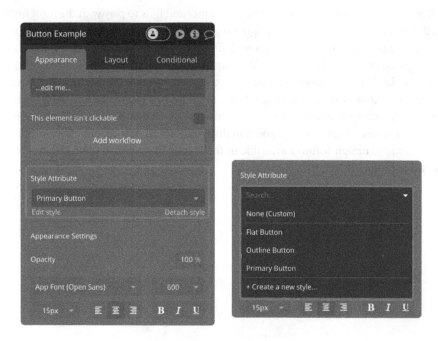

Figure 5.18: Choosing a style attribute

Another tip is to create a dedicated page inside your Bubble project that lists all the styles you have available alongside examples of buttons, texts, groups, and more using the styles you've created. This is called a style guide; search the web for more information about it to get familiar with this concept. By creating such a page, you will have a visual library of all your components and styles, which will make it easier for you to find them, see them together, and even copy and paste them while building. This can save you a lot of time while you're developing new pages.

As you can see, styles play a huge part in your application-building process while using Bubble. They are very powerful and useful, and I am sure that when dominating this area, you will be able to build beautiful projects with a tight design that not only looks cool and is easy to maintain but also helps users navigate and use your product, providing them with a great user experience. Make sure you spend some time playing with components and styles and the property editor to understand all the possibilities and settings you have available to tweak the native components and make them match your brand colors and design needs. I'm sure you will be able to make them look amazing – just keep practicing and learning as you go.

Summary

Layouts are the foundation of any page, and styles can give life to your projects. Mastering both areas inside Bubble is the recipe for a successful project that considers the best design and user experience techniques, making it a pleasant experience for the users navigating your product.

In this chapter, you learned valuable lessons and techniques and how to play with the most important areas of the Bubble editor so that you can configure your layouts. You also learned about grids and how every setting can expose new options for you to define your groups and other elements exactly the way you want, giving you the flexibility to build pretty much anything with no code visually. You also learned how to style elements, where to locate the settings you need to customize every component available, create new styles, edit and manage them, and how you can create and use variables that connect with styles to help you save time and stay more organized and consistent while designing inside Bubble. In the next chapter, we are going to dive deeper into how to build user interfaces and explore the responsive design features available in the responsive editor. I hope you are excited to continue your learning journey!

6

Building User Interfaces with Bubble

In the previous chapter, you learned how layouts and styles work. In this chapter, you will learn about **user interfaces** (**UIs**) and how to build layouts that adapt seamlessly to different devices and screen sizes. We will explore the responsive design features available in the responsive editor, and with this knowledge, you will be able to create pages and UIs that are not only visually appealing but also intuitive and user friendly. Making your layouts responsive and properly configured will allow them to adapt across various devices and screen sizes.

In this chapter, you will learn about the following topics:

- How to create responsive layouts
- How to use responsive design features inside Bubble

Getting started with responsive layouts

Inside Bubble, you can build various layouts and pages for your application, but they are not naturally responsive by default, so it is important to make sure you configure it to behave correctly on different screen sizes. First, start by building your desired layout and then you can start making changes to it to make it adapt and behave in a responsive way. Making a layout responsive will depend on how you set up specific container components and the settings you apply to it directly, using the knowledge acquired in the previous chapter, as well as some other settings that are available inside the responsive mode. Let's learn a little bit about it.

What is responsive design?

Responsive design refers to a design approach used in creating websites or applications that automatically adapt and adjust their layout, content, and features to fit various screen sizes and devices.

An example of different devices and screen sizes are shown here:

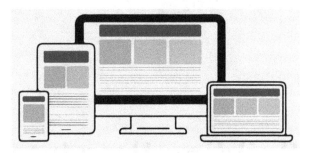

Figure 6.1: Responsive design illustration

It ensures that the user experience remains optimal, whether someone accesses the site on a desktop computer, tablet, smartphone, or another device. In essence, responsive design enables a website or application to look good and function well regardless of the device being used, providing a seamless and user-friendly experience for everyone.

What are breakpoints?

Breakpoints in web design refer to specific screen widths where the layout of a website or app changes to better fit different devices. They're like markers set to adjust how the content looks and behaves based on the size of the screen.

An example of different devices and breakpoint sizes is shown here:

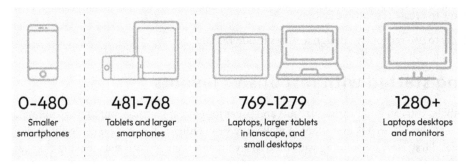

Figure 6.2: Breakpoint sizes example

When the screen reaches a certain width, the design might shift, rearrange, or hide elements to ensure everything looks good and is easy to use on devices such as phones, tablets, and computers.

Now that you are familiar with these terms, let's dive into Bubble.

The Responsive Design tab

When working on the Design tab, the canvas, usually, you are working on the **UI Builder** tab, choosing components, and adding them to the screen. Now, if you want to check what your layout looks like in different screen sizes, you can do this by simply selecting the **Responsive** tab, which is also called **Responsive mode**.

The Bubble interface inside the **Responsive** tab is shown here:

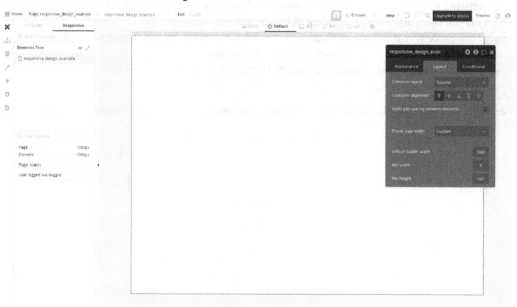

Figure 6.3: Responsive Mode in Bubble.io

Notice that the responsive mode is very similar to the previous mode, but with a few differences – let's break it down. In this mode, we still have the **Elements Tree** section, which works the same way as you've learned before. We also have a list of elements added to the page and their current size. We have a new section called **Page states**, where we can see conditional states in our application so we can test different view modes. One that comes by default is **User logged out (toggle)** so you can view the layout in different ways with a single click; this can serve, for instance, for you to see how a menu changes and how the responsiveness is being affected.

Another thing we can clearly see right away is the new top bar with a few tabs; this top bar is where we can change the canvas size to simulate a smaller screen size and see how our components will react based on the changes. These little options are called breakpoints, meaning they represent different types of devices, such as a larger desktop window (1200 px), a default desktop window (1080 px), a tablet (992 px), mobile in landscape (768 px), or a mobile phone in portrait (320 px).

> **Note**
>
> If, when entering the **Responsive** tab for the first time, you see a message at the top of the page saying **Currently in a fixed layout** and a **Make Responsive** link, you can click that link to change the page layout mode from **Fixed** to either **Align to parent**, **Column**, or **Row**. Selecting one of these options will make your page layout use the responsive mode instead of the fixed layout format, which is not responsive, as explained in previous chapters. In most cases, **Column** as **Container layout** is going to be used to stack sections on top of each other, using Group components as container sections. If you are building a dashboard or a layout with sidebars, then maybe using **Row** would be ideal. **Align to parent** is also an option but for particular use cases. Once you select one of these layout modes, the message will stop showing and the responsive mode will be active, showing you the breakpoint bar with options to visualize your page under the responsive viewer like you are learning how to do in this chapter.

An example of the breakpoints available and options to edit are shown in the following image:

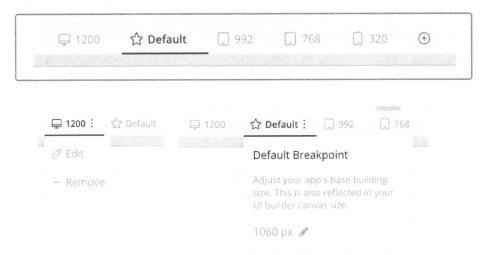

Figure 6.4: Bubble.io Responsive Mode Breakpoints

Using the breakpoints bar, you can quickly change the page layout container size and visualize how your layout will look in a specific device size. You can also create custom breakpoints with specific sizes if necessary. See how you can do it in the following image example.

Here is an example of adding custom breakpoints:

Figure 6.5: Bubble.io Responsive Mode – adding custom breakpoints

Breakpoints are a faster way to check how your page layout looks in different sizes. These sizes are an industry standard, but you can also customize them and create your own breakpoints, adding more or even less. It is all related to how you want to work inside Bubble and how your design and grid should work. If you have design skills, feel free to change them, but if you don't, maybe stick with these options for now.

The **Default** breakpoint is your standard page layout size – it is the same size you view in your normal editing mode, outside of the **Responsive** tab. By default, it is 1080 px, but you can also change it if needed. To change from one breakpoint to another, simply click the tabs and you will see your canvas size changing. You will also notice a little handlebar at the sides of your canvas – you can click there and drag to increase the page size or reduce it. To add a new one, click the + button, and to edit an existing breakpoint, click the ... icon next to the breakpoint size; here, you can change an existing breakpoint size or even delete it completely.

Now, with your layout ready, once you go into the responsive mode, you can click on the breakpoints. Pay attention to what happens – is there anything breaking? Probably yes, and that is OK! Going inside the responsive mode is exactly what we want to do to spot what is going wrong and fix it. When you are inside the responsive mode, you can still click the components, see the property editor, and make changes to your components, which will help you fix them. The secret is to first build your layout and then use the responsive editor later to fix the issues; this can save you time because you can do it all at once. Also, bear in mind that sometimes there might be a problem with your layout structure that might be breaking things, so make sure to fix your layout structure first instead of changing settings. Sometimes, there is no setting that will fix a structural problem, and only rebuilding some sections of your layout will do, so keep this in mind while building your layouts.

An example of a layout inside the Responsive Mode at the **Default** breakpoint is shown here:

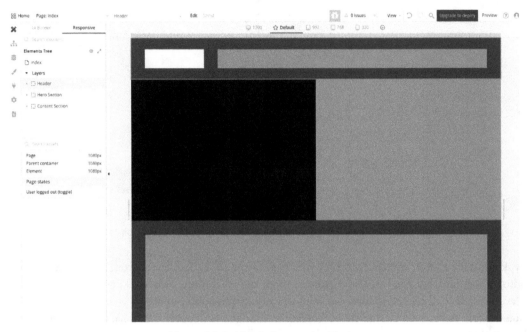

Figure 6.6: Bubble.io Responsive Mode

Be careful with margins and minimum widths. When you reduce your page size, if a component has margins or paddings, it will count as an extra value apart from the actual element size. So, make sure to not set minimum widths if the screen is smaller than the component so that it can adapt accordingly. If your screen size is smaller than the component minimum width, it will not shrink, but it will discount the margins to try to reduce the gap, which can break your page due to this element not fitting in the existing page space.

An example of a layout inside the Responsive Mode at different breakpoints is shown here:

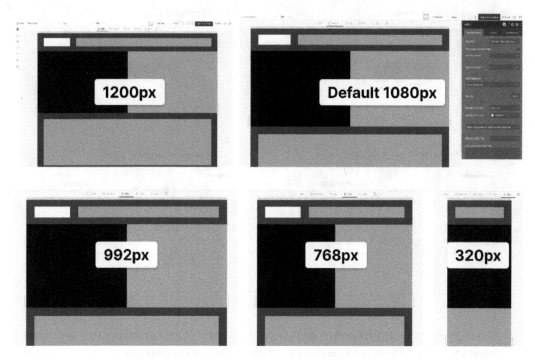

Figure 6.7: Bubble.io Responsive Mode with multiple breakpoints

Once you click on your components, you will want to go to the **Layout** tab to see what options you have available and test which options might help you make the layout more fluid. Usually, there are a few tricks that you can use to do this. Here are a few settings that will serve well for almost any layout:

- Use groups to create bigger sections or boxes that contain other boxes on your layouts. Set this container as **Column** or **Row**. Use **Column** if you are going to add components inside it that will be stacked on top of each other, and use **Row** if you are going to add components inside it that will be side by side.

  This Bubble website layout example uses containers as boxes to build the layout structure:

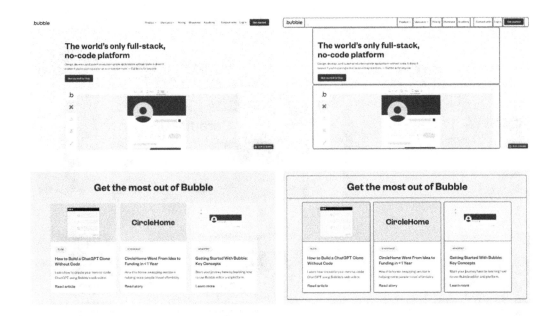

Figure 6.8: Layout structure examples

Notice how a page layout consists of boxes inside other boxes. Remember this concept: a container is a box. The page is the first parent container that contains other boxes, and all the boxes inside the container are its children.

An example of a parent-children relationship is shown in the following image:

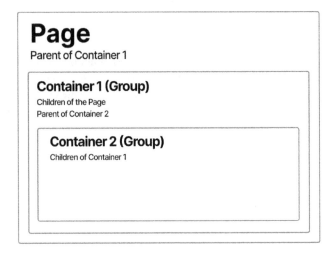

Figure 6.9: Parent and children relationship

Now that you understand this concept, let's move to the second step, which is also very important.

- Make sure the parent component is set properly; usually, it will be the page or the component that is wrapping it. If it is the page, it is most likely to be set as **Column** because a page usually has a lot of Group components on top of each other, forming big sections. In most cases, it will be either a **Column** or **Row**.

An example of different layout configurations, **Column** and **Row**, is shown in the following image:

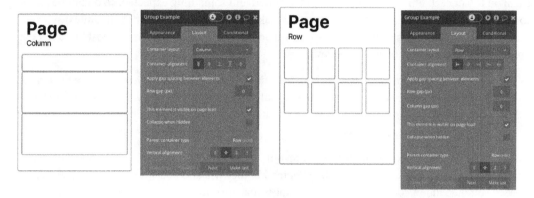

Figure 6.10: Layout examples

- Once you decide whether your group will have a **Column** or **Row** layout, define the container alignment. Usually, it is going to be centered, but you can choose other options, too. Make sure your component is *not* a fixed width – uncheck this option. Leave the maximum width as infinite, unless you want to control how much it will grow, and the minimum width can be as low as your smaller breakpoint. If it is 320 px, also make it 320 px so your page stops shrinking once it hits this size. If you add margins or paddings, remove that amount from the component size so it can fit properly. Choose the horizontal alignment, set the height if needed, or let it grow with the content.

The key here is to test and adapt. If you click to visit the parent component and change it or just tweak the children, you will notice how there is a very close relationship between both components and their settings. Sometimes, what you need to change isn't on the component you are working on, but on its parent, and sometimes, it is not on the parent itself, but on the children. If you have multiple similar components, it is wise to add the same style to them, in case they look and behave the same. This will save you time while making changes to your layout structure, so you don't have to change each individual component independently.

That's it! If you follow these rules, you are 90% good to go, in most cases. The key concept to have in mind is this: make sure your components are fluid, not rigid with specific sizes, so they will adapt as the layout grows or shrinks. By not setting fixed widths or heights and letting it grow/shrink, you will have a component that will be elastic and able to adapt to most scenarios without a headache.

> **Tip**
> To separate or define spaces between components that are close together, use margins or paddings. Think of your layouts as square boxes; this will help you organize and understand them better.

Notice that it is very normal to have components pushing each other to the bottom, so keep in mind that when a layout adapts to fit a smaller screen size, it is very normal to "break" your sections and pile them on top of each other. This is a common standard and something expected, so do not be afraid to let some blocks stay on top of each other and make the layout change a little bit while displaying it on smaller screens. Naturally, a wider layout will become narrow and become almost like a strip of components when visible on a mobile device.

When you notice a component is not working well on a specific breakpoint, you also have the option to hide it; there are some cases where not all the information available on a desktop device is also needed on a mobile screen size, and sometimes you will need to not only hide a specific section or component but also change others completely.

Be careful: if you do make changes to your layout while working on different breakpoints, these changes will not only affect that specific breakpoint, but they will also affect the project as a whole. So, how do you then hide a component or apply changes that are only going to be visible in a specific size without affecting the others? The answer is by using **conditionals**. When you click a component and go to the **Conditional** tabs, you can simply add a new one. If you start typing its name, you will see a little drop-down window showing a few options. Locate options related to the breakpoint size, and add your rules using the size you want to target. For instance, you can choose to make that specific component a different color or a different size, or even choose to not display it completely. If you choose to not show it, also make sure you check the **Collapse when hidden** option, which is available under the **Layout** tab. This will make other components ignore the components that are not visible in a specific situation and just occupy the newly available space. This trick can solve a lot of layout issues!

Here is an example of a **Conditional** tab showing breakpoints that are available to select as a conditional rule:

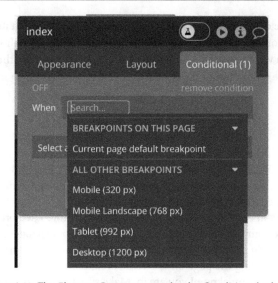

Figure 6.11: The Element Property panel – the Conditional tab rules

The list of breakpoints can vary according to the breakpoints available on your project. As mentioned before, you can customize it to your needs.

To make a component invisible in a specific breakpoint, follow this configuration:

Under **Conditional**, click the **Define another condition** button, click **When**, choose the **Current page width** option, choose an option (e.g., **> Smaller than** or **< Bigger than**), breakpoint (e.g., **Mobile Landscape 768 px**), and then click outside of your selection. Now, under **Select a property to change when true**, choose **This element is visible**, and uncheck the checkbox.

Here is an example of a conditional rule used to hide the element on a specific breakpoint:

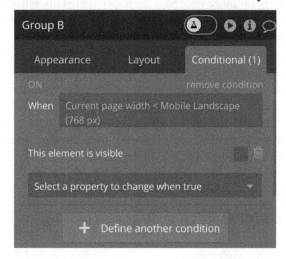

Figure 6.12: The Conditional tab with a rule applied

Make sure to also click the component and check it as **Collapse when hidden** under the **Layout** tab. Now, change your breakpoints and notice how this component will hide or show depending on the breakpoint selected. If you want to add an animation to it, check the **Animate the collapse operation** box and choose one of the animation options available. This option can help make the transition a bit smoother.

> **Note**
>
> If you are familiar with other no-code tools or have used some before, it is important to note that breakpoints inside Bubble may work differently. In some no-code tools, such as Webflow, if you are editing your page while viewing it in a specific breakpoint, changes applied to the layout will only be applied to that specific breakpoint. That is not how it happens inside Bubble. Consider your page layout and the breakpoints as one single thing. Each breakpoint is only a way to quickly visualize your page on that specific window size, but changes are global and will affect all the breakpoints. So, keep that in mind while editing your layouts inside the responsive editor. The fact you are editing the page on a specific breakpoint does not mean that changes are recorded based on which breakpoint you are in. Make sure to test your changes accordingly to avoid fixing one specific break size but causing problems when you change it to another one.

Elements Tree

Another important area inside the **Responsive** tab is the **Elements Tree** view. We covered this area in *Chapter 2*, but now we are going to talk about it while you are inside the **Responsive** tab. As you may have already noticed, **Elements Tree** keeps showing up if you are under the **UI** tab or the **Responsive** tab. The reason for this is that you are going to continue using it to locate components on your layout and interact with them as needed. When working on the Responsive tab, the Elements Tree panel will behave the same way. The difference is that, under the **Responsive** tab, your canvas will have breakpoints and your goal is going to be to optimize your page for responsivity rather than building your layout from scratch. Since you don't have the UI elements area visible while in the **Responsive** tab, it is not as practical to add new components while in the **Responsive** tab. So, this will naturally require you to keep switching from the **UI Builder** tab to the **Responsive** tab from time to time. Once you click a specific element under the **Elements Tree** tab, you will notice it will show a little preview of that component on the side, and the name of that component will become highlighted. If there is an arrow on the side of the item's name, it means it is a Group component and it contains other components inside it. You can click the arrow to expose the components inside that `Group` component and see the whole tree structure of components. If you wish to see the entire tree structure at once, you can click the little icon (with two arrows in opposite directions) at the right side of the **Elements Tree** section name and expand or collapse the entire tree structure with a single click. Next to it, there is an eye icon that can be used to reveal hidden components if needed.

Here is an example of **Elements Tree** while in the **Responsive** tab:

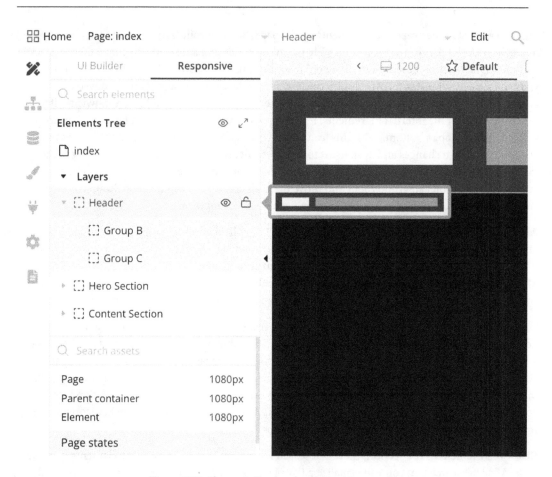

Figure 6.13: Elements Tree under Responsive mode

While working on the **Responsive** tab, **Elements Tree** will help you select specific components to access them and maybe make a few changes to the style to make it work while in a specific component. You might also need to change a section, rearranging its position. To do this, you can simply click the element name under **Elements Tree**, and you will see that it is highlighted on the right side, under the layout viewer (canvas). The opposite can be also done – you can click the component itself on the page and it will highlight it under **Elements Tree**. If the component was hidden before, because it was inside a Group component, for instance, it will open the folder and reveal where it is located. So, it is up to you to find the best approach while working on Bubble. To rearrange an entire section, for instance, you can click a specific element under **Elements Tree** or click the component on the page and then head to the selected element under **Elements Tree** and simply click and drag it to the top or bottom depending on where you want it to go. For instance, if you have a section at the top of the page, which is a Group component with various components inside it, but you want to make it go to the bottom of the page or between other sections, you can click on that Group component, move it closer to the other Group component you desire and drop it there, inside the **Elements Tree** section.

This will rearrange the order of components on the page. Because your layout is responsive, once you move the order of the components under **Elements Tree**, this will automatically make them restack and find their position based on the type of layout options you choose. If you choose **Column** for your entire page, for instance, each Group component can be a section, and if one is moved on top of another, they will change order based on which one comes first. Think of it like a sandwich – if you move the order of the internal ingredients, it will change which one comes first because they are all stacked up from top to bottom. With this feature, and because your layout is responsive, it is easy and fast to make design changes and even use it to test different layout structures without having to move individual components one by one; you can simply move an entire group or section. If components are not grouped, you can group them and move them all and maybe remove the Group component later if not needed. This is also a great tip to help you make changes faster.

Another way to change the order of the components is by clicking on them and going to the **Layout** tab. There is a particular area called **Parent container type**, which will usually be **Column** or **Row**. Below it, there is either a place called **Vertical** or **Horizontal** alignment. Below this are a few buttons: **Make first**, **Previous**, **Next**, and **Make last**. If you use these buttons, you can easily move the selected component across the layout structure. Give it a try and you will understand how it works.

Under **Elements Tree**, you can also double-click a specific component name and rename it easily. You can also toggle it on or off to make it visible or invisible, which can help you visualize what will happen with your page layout if you hide something temporarily. This can be a nice way to test components that change based on a specific logic of page state, for example, once you have logged in menus and buttons that change depending on if users are logged in or not. Another option is to lock components, in case you don't want to mess something up – you can click the locker icon and just make them non-draggable.

Below the **Elements Tree** section, there is a search bar that can be used to locate elements on your page. After the search bar, you will visualize a few pieces of information that are only available under the **Responsive** tab, such as the page size, which is based on the breakpoint you are in, and the parent container size, which is the size of the parent Group component or page holding the component – it is usually the page size. These numbers can be used to compare and see whether settings are matching and properly configured. The last piece of information is the element size, which is directly related to the element selected now; if you select new components, the value will change accordingly. A good tip is to check whether specific element sizes do not surpass the actual parent container size or page size.

Below that area, we can see an option called **User logged out (toggle)**. This feature can help you visualize changes related to the user being logged in or out. For instance, you have a menu that has a condition that changes its visibility based on the fact a user is logged in. When you click that option, it can trigger the condition, allowing you to visualize, under the responsive mode, what these changes will do to your page layout. So, basically, you can use this option to simulate that the user is either logged in or out and see how your layout will change based on that. If you do not use any of these resources, then you don't need to worry about them or use them, but just know why they exist for now.

Suppose you need more room to preview your layout while in the responsive mode, you can click the little arrow on the very edge of the sidebar and collapse the sidebar entirely. That can be a great way to give your browser a little more room to test and preview the responsiveness of your page layout.

Now that you've learned how the **Responsive** tab works, let's talk about how you can preview your responsive layout in other ways and how to use different methods to test it and make sure everything is working fine.

Different ways of previewing your pages

Using the responsive mode is one way of visualizing how your layouts looks in different screen sizes, but it is not the only way. In this section, we are going to take a look at other methods for testing your application in different screen sizes.

Preview mode

As you may already know at this point there is a button called **Preview**, available at the top bar of your Bubble Editor. If you click it, a new window will open and load the current page you are working on in the editor. With the **Preview** mode, you can quickly navigate to your page, and by stretching the browser window size, making it smaller or bigger, you can have a sense of how your layout is adapting to the changes.

Browser Inspector

Another way to validate how your page layout looks using your browser is by activating the **Browser Inspector** feature. After clicking the **Preview** button, once you are on your desired page, hit the shortcuts on your keyboard, which are *command + shift + C* (on a Mac) or *Ctrl + Shift + C* (on a PC), and a new window will open on the right side, showing some code. In this same window, at the top left, locate a little icon showing a laptop and a mobile device. This will allow you to enter the device simulation mode. You will notice your browser will change – at the top, you will visualize different device names; you can change them to test your page on different device types and sizes as if you were actually using them.

The Chrome Browser Inspector mode shows the responsive preview as shown in the following image:

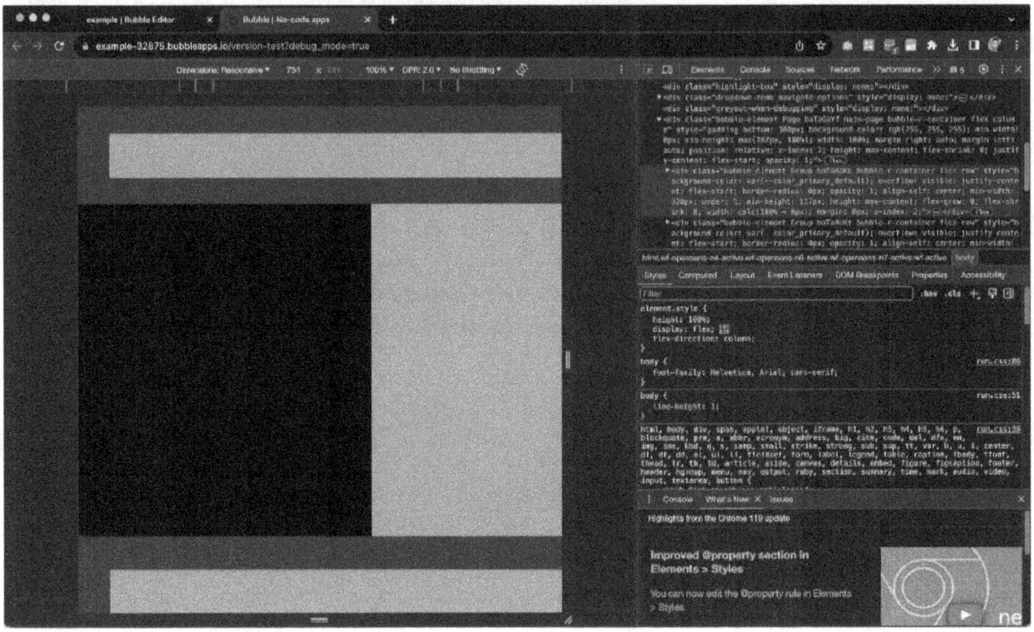

Figure 6.14: Google Chrome Browser Inspector

> **Note**
>
> This works on Google Chrome, but other browsers are different. To learn more about the Google Chrome inspector, you can search for specific tutorials on how to better navigate and use it. There are a lot of other features and things you can do with it, and it can be helpful to know how to use it to debug your applications.

Real devices

Another way of testing your layouts is by actually opening pages on different devices. This is the most accurate test you can do, but, of course, it's not the most effective because you are not going to have a lot of different devices to test in real life. This is why we usually do tests in a virtual way, as shown before. However, you have your desktop computer or laptop and a mobile device, so you can use them to also test and see whether everything is working properly. To do that, simply go to the **Preview** mode, get the link of your application on test mode, and share it across your devices, so you can open it inside your mobile phone browser window, for instance. Nothing beats testing the user experience on the actual device, so, try to test it for real whenever possible, and if not possible, use virtual options to help you speed up your development process.

Websites to test multiple breakpoints

There are a few tools and resources online that can help you test your page in multiple breakpoints at the same time, which can be a good way to simply check whether your page is adapting well in various scenarios. Here is one free and open-source tool you can use to test your layouts in multiple sizes without much effort: `https://responsively.app/`.

With different methods of testing your page layouts, you will be able to make it work perfectly across different device types; now, you have a lot of options to choose from and can become a master at creating amazing responsive designs.

Building responsive designs at first can be tricky, but that's OK – the main idea here is to get familiar with the key concepts and start playing with them. Once you do this a few times, it will become more familiar and natural and it won't be a problem anymore.

Keep testing and building layouts and use tools that can help you along the way. A good idea is also to try to learn from well-known and famous websites so you can get an idea of how they were built and learn from real-world examples. I hope you become a responsive master in no time!

Summary

In this chapter, we've covered the essentials of the responsive editor. You learned about responsive design and breakpoints and how to configure them inside Bubble.

You also learned important tips and tricks on how to properly set up your responsive designs and important concepts that will help you understand how to work with responsive layouts while building your no-code applications.

Understanding these concepts will be important for any future project you build. In the next chapter, you are going to learn how to use workflows to give life to your applications.

7

Workflow Automation and Logic

This chapter will introduce you to **workflows** and will teach you how to configure elements and create **actions**, **triggers**, and set up **logic statements**. We will dive into the **Workflow** tab and learn how workflows work and what you can do with workflows. You will also learn how to integrate your designs (the frontend) and the database with workflows to allow you to create powerful applications. Workflows are an essential part of any Bubble application while the database is the heart of your product, the frontend is the face of your application, and the workflows are the brain, so, you need to know how they work together.

Additionally, we will also explore backend workflows and explain how to use them to allow you to build cool automation inside your projects.

In this chapter, we will cover the following:

- An introduction to workflows and logic features
- Creating actions and triggers for user interactions
- Managing workflows and elements
- Incorporating conditions and logic statements
- Backend workflows

An introduction to workflows and logic features

What are workflows and logic triggers? Workflows are a sequence of actions or steps. Think of it as a recipe or a checklist, a sequence of steps to accomplish something. This sequence will be activated by specific events or user interactions that happen inside your application, which we call triggers. Logic triggers fire the sequence of steps configured inside a workflow, and then things happen.

For instance, when a user clicks a button, it creates a new user account and sends them to another page. It might sound simple, but for that to happen, multiple little steps had to happen in order for that account to be created. The trigger is the person clicking the button and the workflow is the little steps that happen to create the user account, store the information on the database, and then redirect the user to another page.

The cool thing about using Bubble and no-code is that you will be able to define workflows and triggers using plain English without writing code. How cool is that?

Now that you know workflows are like a set of instructions, such as a recipe or a checklist, you understand the concept. However, it is important to understand the difference between how we as humans work and how Bubble works. How do workflows work? If I give you a recipe and ask you to cook it, you will probably be able to just get things and start making it. Now, with Bubble, we have to be a little more careful. Think about it like this: we, as humans, know our surroundings, objects, and everything around us, so we can quickly adapt and do things, but Bubble is not a human – it is a piece of software. So, we need to tell it what to do, but also, everything else it needs to know. So, for instance, if the first step is to get a specific ingredient in the fridge, you are a person who knows what is a fridge, where it is, how to open it, where the ingredient is, what the ingredient is, how to get it, how to close the fridge door, and so on. Now, when dealing with workflows in Bubble, you will have to specify every little step and information it needs to know to perform a specific action, so, you will have to give specific instructions to it. For instance, step one: open the fridge (name, location), step two: locate the ingredient (name), step three: get it and put (location), step four: close the door (item name, location), and so on. As long as you understand this concept, you will be fine and be able to create amazing workflows. At first, you will probably forget a few steps and definitions, because we still think Bubble will know what we mean, but it won't unless we specify it very clearly, and after a few trials and errors, we will get used to providing information differently and creating better recipes that it can follow for us.

Creating actions and triggers for user interactions

Let's dive in and get familiar with this very powerful area inside Bubble. In this section, you are going to learn how to create workflows in practice. Get familiar with the **Workflow** tab – you are going to use it a lot.

The **Workflow** tab inside Bubble is shown here:

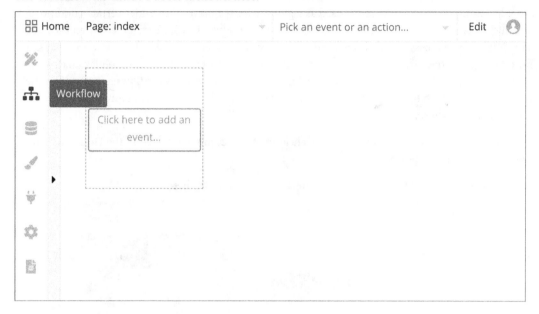

Figure 7.1: The Workflow tab

There are two ways of creating new workflows:

The first way is by going to the **Workflow** tab and clicking to add an event. You will have to pick an option and then specify which component or action will trigger the workflow.

Workflow categories and actions are shown here:

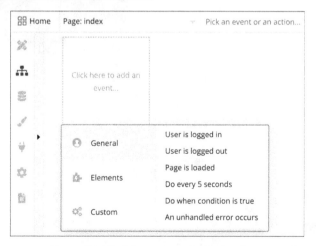

Figure 7.2: Workflow categories and actions

If you create a new workflow without choosing an element, you will have access to a few options under the **General** category. It can be used as a global trigger that will fire under certain conditions, meaning no one needs to click anything to start that workflow – it will be triggered when the page loads. You can also access custom events and create them from there – there will be more about this in the following sections. If you choose the **Elements** category actions, you will have to pick an existing component available on the page you are working on. Remember that workflows are related to the pages, so if you switch to another page you won't see the workflows available before. Each page has its own set of workflows.

The second way to add a new workflow is by clicking on the component itself in the Design tab and exposing the **Property Editor** panel.

An example of a selected button and the **Properties** panel is shown here:

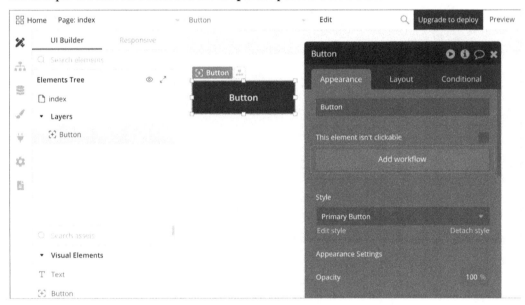

Figure 7.3: Properties – Add workflow

Once the component is selected, the Property Editor will be visible so you can click the **Add workflow** button to create a new workflow associated with that component. You will be redirected to the **Workflow** tab with a new workflow ready to be created for that specific component.

An example of a new workflow created from a button element is shown here:

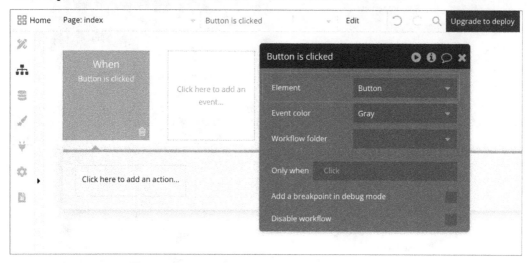

Figure 7.4: New workflow when Button is clicked

As you can see, this new workflow uses the selected element as a trigger, meaning it will run every time someone clicks that button. In this example, it is a Button component, but workflows can be added to multiple types of components. If you want to add a workflow to a component, there is no **Add workflow** button under the **Properties** panel, but you can still do it. You can simply right-click the component and choose the **Add workflow** option or hit the keyboard keys *command + K* (Mac) or *Ctrl + K* (PC).

Here is an example of an input field that doesn't have the **Add workflow** button:

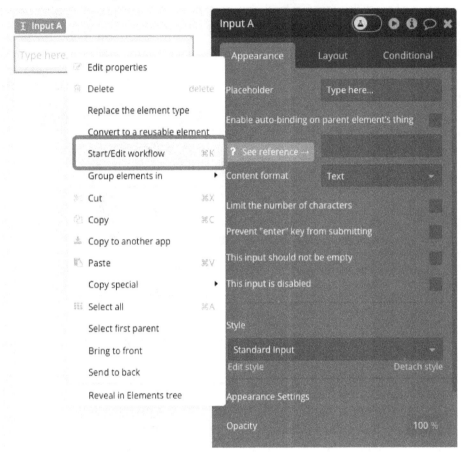

Figure 7.5: Add workflow when a button is not available

Now, once you are inside the workflow, you can start creating the automation you want to run – this is the logic part inside Bubble. Think of a workflow as a process – like a factory assembly line, it is a group of actions, and you can choose how many actions are going to be running at that specific moment. In this example, once someone clicks a button, you are free to choose – it can be one single action or multiple running at the same time. It is like automation. Think of cause and effect – the cause is what makes it fire, and the effect is the little actions that will happen when the workflow starts running. Now, this is an important concept to keep in mind: since this is a trigger, the actions run all at the same time, asynchronously, meaning that once the workflow starts, it will run each individual action altogether; it won't wait for one to complete before running the next, unless you explicitly specify it. Don't worry too much for now – just know that this is how it works. It is just important to understand the concept, so if you ever need one action to run in a specific order, you can add delays to force the workflow to run one action before another.

Now, let's dive into the types of little actions you can add to your workflow; there are lots of possibilities. The actions are divided into groups – let's take a look.

The example of multiple categories and workflow actions is shown here:

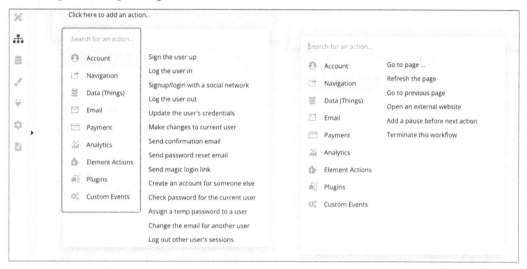

Figure 7.6: Workflow actions

Once you click to add an action, Bubble will expose a list of categories you can choose to find the specific action you are looking for. If you are not sure, you can also use the search bar at the top of the list and type the name of your desired action, as this will help you filter existing options and find them quicker. The first category is **Account**, which has all the actions related to handling user creation, login, sign-up, and so on. Inside this category, you will find all the actions available to let you manage and create features that are related to user accounts.

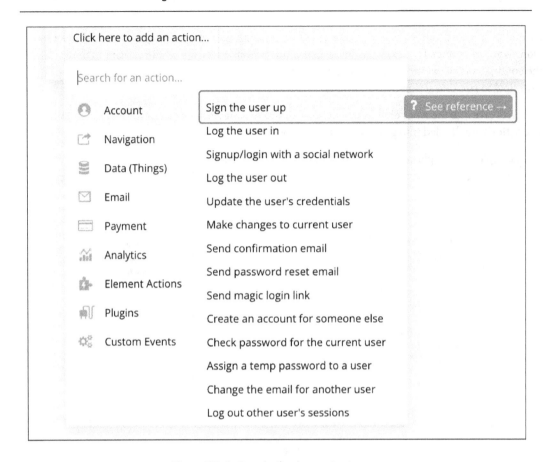

Figure 7.7: Actions in the Account category

Note that it might be overwhelming at first to understand what each action does and where to find the one you need to use. To help with this, use the search bar to locate specific actions. Another piece of advice is to hover your cursor on top of the action and a little tooltip will appear saying **See Reference**; if you click that link, it will send you to the Bubble official documentation page, where you can learn a little bit more about what that action does and how it works. Another piece of advice is to simply use it and test how it works to learn in practice.

The next category is **Navigation**, where you will find actions that will help you build links between pages, such as the **Go to page...** action and some other actions that allow you to let your users navigate through your application.

Another very important category is **Data (Things)**, where you will find actions that allow you to create, remove, update, and delete items on your database. This is what we call **CRUD** in the development world – it is a very helpful set of actions that will enable building dynamic apps and handling database operations.

Once you locate the action you want to use for your workflow, all you have to do is click on the chosen action and it will be added to your current workflow. It is very simple: choose an element, create a new workflow, choose the action, add it to the workflow, and repeat the process over and over, and that is how you create logic inside Bubble, using plain English, without code!

A workflow with the first step being to sign the user up is shown here:

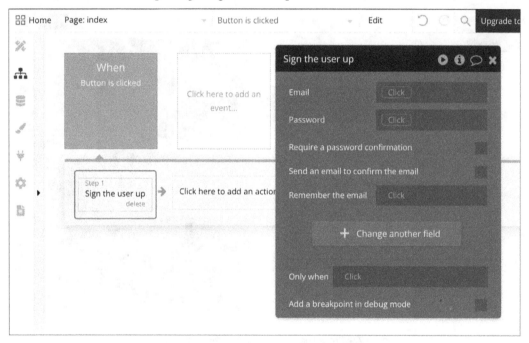

Figure 7.8: Workflow Step 1 – Sign the user up

As you can see the **When Button is clicked** workflow has a first step, which is the **Sign the user up** step. Now, this will run every time this button is clicked, and it is very straightforward to understand. Once you click on the action, it will expose the **Properties p**anel, allowing you to configure how that step should work. Each action will have its own set of settings and features for you to configure and steps can also connect with each other, sharing data and information across your application.

If you exit the **Workflow** tab and click on an existing component again, you can see that the button has now changed to **Edit workflow**. If you click it, it will go back to the workflow for you to continue editing it.

The **Properties** panel shows the **Edit workflows** button:

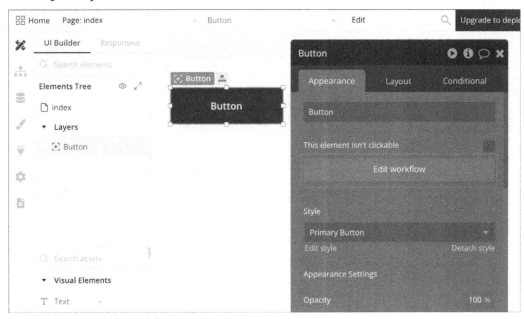

Figure 7.9: Properties – the Edit Workflows button

Let's take a look at the next actions and categories available. You will also find actions to allow you to send email messages and actions for payments and analytics. As you can see, payments and analytics are not available by default, but you can find amazing plugins to install and enable actions for these two categories. Once you install a new plugin that has workflow actions, the action names will appear under these categories, and all you have to do is pick them. A common plugin for payments is **Stripe** or **Paypal**, and for analytics, you can use **Google Analytics** and **Mixpanel**, but these are just examples; there are a lot of options available for you to explore, so you can take a look at the plugins marketplace and have some fun!

An example of the categories and workflow actions available are shown here:

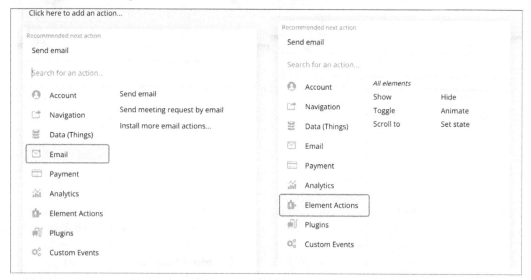

Figure 7.10: Workflow actions and categories

Other very commonly used actions are **Element Actions**, which allow you to manipulate how elements behave on the frontend of your application. You will find options to show, hide, toggle, and even animate components, and a cool thing is that it helps you manipulate what we call **custom states**, which are a very useful and interesting technique you can use to define conditional variables on your components. We are going to talk more about that later.

An example of more available categories and workflow actions is shown here:

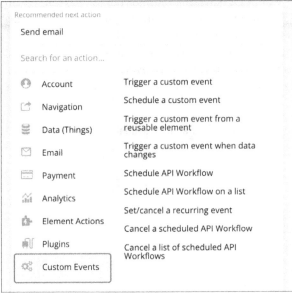

Figure 7.11: Workflow actions and categories

Last but not least are the **Plugins** category and **Custom Events**. Like payments and analytics, the **Plugins** category will show you all actions available related to the plugins you have installed on your Bubble application. Not all plugins will bring new actions to be used, but some will. So, if you install a plugin that has actions, this is where it will be visible. **Custom Events** is an interesting part of Bubble; it will allow you to build what we call **backend workflows**. You can choose to create custom workflows that are not necessarily associated with a particular component, but that can be called into a workflow at any moment and by multiple workflows, which can be a way to create repeating actions that essentially work the same but sometimes only change one or two variables. It is a bit more complex, but very useful. To simplify, think of it like a cake recipe you can call a robot to execute at a specific time, a custom event is a type of recipe that can combine multiple actions, like a simple workflow, but that recipe can be called in the middle of any workflow or scheduled to run on the background on a specific date and time. This allows you to create more complex types of applications, for instance, if a person clicks on a button, you can just program a custom event to run later in the future, which can be used in multiple scenarios and is a very powerful feature inside Bubble. Take some time to play with it and get more familiar with this feature.

Managing workflows and elements

As you can see, each page will have a few workflows, so it is very common to have a huge list of workflows in no time. To help organize workflows, Bubble has a few features that can make your life less chaotic when building and using workflows.

An example of settings to organize workflows is shown here:

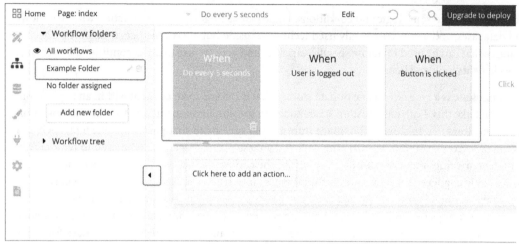

Figure 7.12: Workflow organization

If you click on the little arrow on the left, close to the sidebar, it will open more options. On the workflow **Properties** panel, you can choose colors and create folders to better organize your workflows. You can choose, for instance, to group them using folders to specify what is what. You can also assign colors to specific types of workflows to be able to better identify for instance, which workflows are actions, which workflows are dealing with errors, which workflows are related to payment, and so on. You can create your own organization system using these features.

Workflows and logic play an essential part in building applications with no-code and although the basic concepts are simple, they can get really complex as well, which can give you an idea of how powerful no-code tools can be. My advice to you is to start simple and avoid complex workflows and crazy ideas; if it sounds too difficult, chances are, you are making it more difficult than it should be – take a pause, let it sink for a while, then come back and play with it again, and allow yourself to explore new ideas. In the beginning, you will get stuck eventually, but this is part of your learning process; allow yourself to learn and keep going. Since workflows and logic are complex by nature, it will require some time and effort to master. However, be thankful that it is way simpler than writing code; after all, you are building everything with plain English rules and conditionals.

Incorporating conditions and logic statements

Now, let's dive into some logic and conditions. Logic plays a huge part in conjunction with workflows and databases, and a very commonly used technique inside Bubble is called **custom states**. This is a resource that can be used to define specific variables and you can use them as conditionals to change how elements look and behave.

A custom state is a type of variable or data storage that is unique to an element within an application. Think of it like this: I can have custom states added to specific components and use that variable from it, which allows you to store and manage information specific to that particular element, such as a button or a group, without affecting the overall database. Custom states can be used to temporarily hold data or values that are relevant only within the context of that element and are typically used for dynamic changes or interactions within the user interface. It is not something you will store on the database, but it is something that can be fetched faster, similar to computer memory. With this resource, you can store and update information related to a single element's behavior or appearance inside itself. This feature can help you create interactive and personalized experiences for users without permanently altering the underlying data structure of the application. Let's see how you can add custom states inside Bubble.

First, locate the component you want to use and click it to reveal the **Properties Editor** panel.

The **Properties** panel example displaying custom states is shown here:

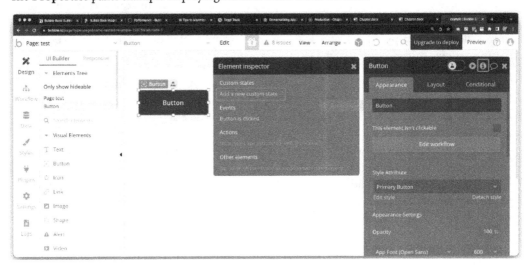

Figure 7.13: Custom states

On the right top side of the **Properties Editor** panel, you will see a little circled **i** icon. Click on it to expose your component's settings and also the custom states. Now, just click to add a new custom state there.

An example of how to create and configure a new custom state is shown here:

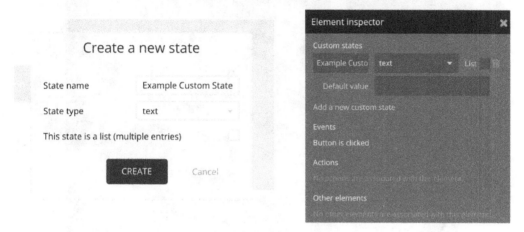

Figure 7.14: Creating and setting up a custom state

A new modal window will open, where you can give your custom state a name, choose the type, and select from the existing options depending on what type of data you want to store; it can be **text**, **number**, or many other options. Once you are ready, click **CREATE**, and now you will see your newly created custom state available on the **Element Inspector** panel next to the Properties Editor panel. That is it! You created your first custom state. Congratulations!

Now, what can you do with it? Well, there are really a lot of things you can do, and it is a resource for you to also explore while building. Now, you can, for example, define a state and change the component appearance based on the current state, so you can use it to change colors and other visual styles, hide or show components or entire sections of a page, and much more. Let's take a look at a practical example. Under the **Conditional** tab, you can, for instance, define a rule, which rules that when the custom state has a specific value, for instance, a Text component called "Example," the button's background color will turn red.

The **Conditional** tab under the Properties panel with custom states is shown here:

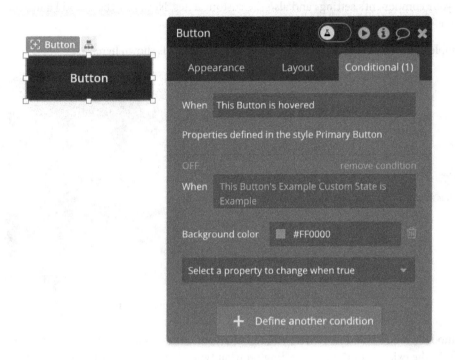

Figure 7.15: The Conditional tab

This logic works like a conditional statement, meaning that things can be changed based on the values inside of the custom state. We can go beyond this and tweak the custom state value using what we learned before about workflows. For instance, we can create an action that will change the custom state value when someone clicks on a specific button or any other component.

An example of a workflow action setting a new value for a custom state is shown here:

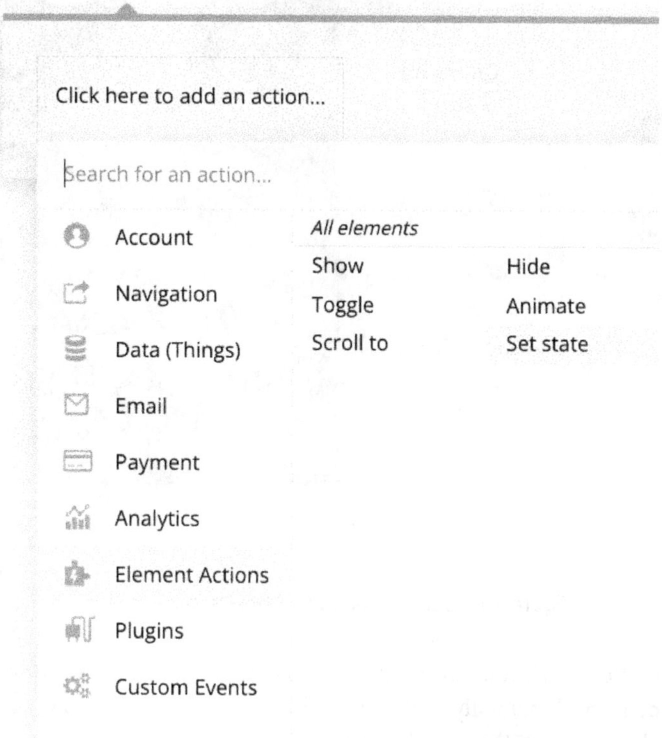

Figure 7.16: Element Actions – Set state

To do that, we can create a new action inside the **Workflow** tab, pick the **Element Actions** category, and add a **Set state** action inside the workflow. This will allow us to define a new value for the custom state we are targeting. Once it changes, the corresponding conditionals will trigger and something will happen based on what we've configured.

The example of a set state workflow action getting a dynamic value from the component state is shown here:

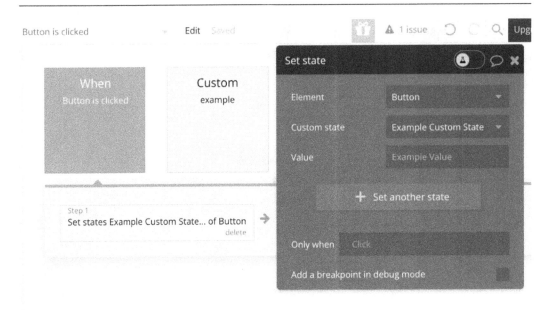

Figure 7.17: Set state action with dynamic value

In the **Set state** properties panel, you can specify which component we are targeting and what custom state we are referring to, and manually or dynamically change the value. Now, with this concept in mind, you can do a lot of things; the most important thing is that you understand its foundation so you can think of various ways you can use it on a daily basis. This logic technique is for sure going to help you build more complex and exciting projects inside Bubble.

Only when

There are other simple features that most people overlook but that can make a huge difference if used correctly: one of these features is the **Only when** filter. This is an option that appears at the bottom of each workflow you set up and can help you define whether that workflow will run or not based on the settings you define.

An example of the **Only when** filter is shown here:

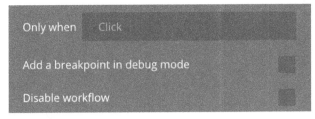

Figure 7.18: The Only when filter

The **Only when** filter can be helpful in many cases, and all you have to do is to write a rule that will make it run or stop it from running. It can be used to create more dynamic workflows that sometimes are running with different steps but not all the time should be performing a few actions, so you can build a single workflow. However, by adding filters, you have better control of what is going to be triggered and what is not.

If you add a rule and the criteria are not met, this step inside your workflow will simply not run, but it will allow the other steps to flow normally until the end of the workflow steps.

The example of the **Only When** filter with a rule applied to only activate for logged-in users is shown here:

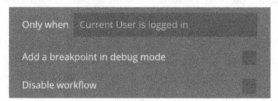

Figure 7.19: The Only when filter with a rule applied to it

A common usage is to filter specific workflow steps based on user access – these rules can help improve the security of your application and even block people from accessing specific pages or preventing sensitive information from loading before the page loads. You should definitely give it a try and get more familiar with this feature.

Workflows run at the same time

As mentioned before, workflow steps run asynchronously, so keep this in mind when building automation. If step one takes more time to load, it won't stop the next steps from running. Think of it like a lot of fireworks being fired at one in a single row – they will all be fired and start going; one won't wait for another unless there is a connection between them and a piece of information is needed from one step to proceed with the other.

To be a little more specific, frontend workflow actions proceed in order. There is a difference between backend and frontend workflows, but subsequent actions do not wait for the previous one to complete before triggering the next one. The actions will run in order, but if there is something that needs to happen that the other step needs to successfully complete, it might fail, because it will not wait for the result of the previous step to run the next.

Now, backend workflows do initiate immediately upon triggering, like the fireworks example mentioned before. So, think of that action as a single action and the steps as each firework being fired at the moment – they are triggered regardless of their position within the workflow sequence. The step number and order in which they were positioned inside the workflow logic you created do not interfere with the outcome just because of the step or whether they come first or last. For instance, even if a **Schedule API Workflow** action is placed last, it will trigger as soon as the workflow is initiated.

This is an important concept to understand to avoid common problems with workflows. These problems can be created unintentionally but frequently, especially depending on how you structure your workflow and logic. For instance, searches may not always reflect immediate updates, especially with new data, which may be added to the same workflow. There will not be enough time for each action to fulfill their configuration, because they are running at the same time, in parallel. While building logic in a sequential order, it is common to think one thing will happen first and then the other second, and so on. Just be careful because this is not exactly how it works sometimes unless you explicitly tangle the results of actions with further steps.

So, be careful if you are relying on search results immediately after creating a new item; it may yield inconsistent outcomes.

Another type of event is the custom event. They execute sequentially, not simultaneously. For instance, if Workflow A triggers a custom event initiating Workflow B, then Workflow B will finish before the remaining actions in Workflow A because it will initiate right when it was requested to start. Just keep that in mind.

By the way, there is an option that allows you to use data generated from a previous step in a further step, which can be useful sometimes.

An example of workflow steps using a result of a previous action/step in the same workflow sequence is shown here:

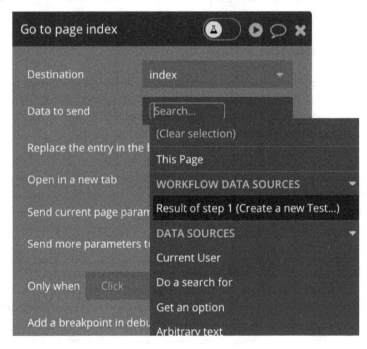

Figure 7.20: Result of step 1

For instance, if you just created some data and added it to the database but you need to also send the user to a new page and use that same information, you can get this from a previous step without having to fetch it directly from the database. This can also prevent errors since steps run individually.

Retrieving an item from the **Result of step X**, where *step X* involves the **Create...** action, is consistently reliable. This can also be a better and more secure way to utilize data from one step to another instead of using searches.

Now there is just one way to delay steps from happening, but this method only works in certain conditions; it is especially related to page navigation. If you go to the **Navigation** category, you will find an action called **Add a pause before next**. This action is how we can a delay step.

An example of a delay/pause step in the workflow sequence is shown here:

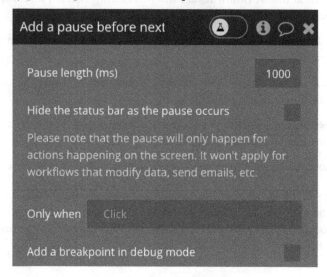

Figure 7.21: Adding a pause step

This action can be used on specific occasions where you want for some reason to delay the transition between pages or slow down some of the steps. Maybe they are running too fast and you need something else to happen before another step starts – this is the only option available to slow down the workflow steps a little bit. To configure it, you just have to add it in the middle of your workflow steps and define how many seconds it will delay; the pause length is added in milliseconds so each 1,000 is 1 second.

As it is written on the feature note itself, this pause will not be applied to all the workflows, but it is an option for some specific use cases. Use it wisely.

As you can see, Bubble does not offer a specific feature for an action to wait for a workflow to be over before moving on to the next step. However, if you really need or want to find ways to connect the logic and the run order of the workflows, you can use some of the examples previously presented in this chapter. To wrap up this section, I will summarize a few tips and extra advice on the same topic, which can help you achieve that:

- Create dependencies between workflow steps by requiring data or information from previous actions

- Group segments of a workflow within another workflow

- Consider creating multiple workflows instead of one if a specific trigger yields multiple results based on conditions

- Place conditions at the workflow level rather than at the action level for better control and management

- Integrate *Step 1* into a custom event if *Step 2* relies on data manipulated in *Step 1* to ensure sequential execution

- Implement backend workflows within custom events positioned after the preceding steps to control their execution order

- Use the **Trigger another workflow** action in the primary workflow to execute secondary workflows before continuing with the main flow

- Insert brief pauses (delays) between steps to facilitate smoother transitions and processing

As you can see, using workflows can be a little complex sometimes, but they are immensely powerful and exciting to use. The more you use it, the better it will be, so make sure to practice. Digging deeper into the workflow world now, let's start learning about the backend features for creating logic and workflows inside Bubble.

Backend workflows

Working with logic and workflows can be very powerful and allow you to build complex types of projects using Bubble; it is amazing what you can do with it. Backend workflows can do a multitude of actions and allow you to run workflows in the background. By nature, this is a little more complex feature so it will require you a bit of time to master it. This subject could be a full book by itself, so here you are going to just learn the basics. It is recommended you keep studying it after reading this book to fully understand backend workflows. Let's dive into it.

The main difference between regular workflows and backend workflows is that backend workflows will run in the background, on the server side of your application, not on the page itself. While using backend workflows, you can trigger API workflows, database events, recurring events (loops), and custom events.

To add a new backend workflow, you first need to configure it under your application settings. Head to the settings and under the **API** tab and locate the **Enable Workflow API** option and backend workflow. Click the checkbox to enable it.

The **API** settings to allow backend workflows are shown here:

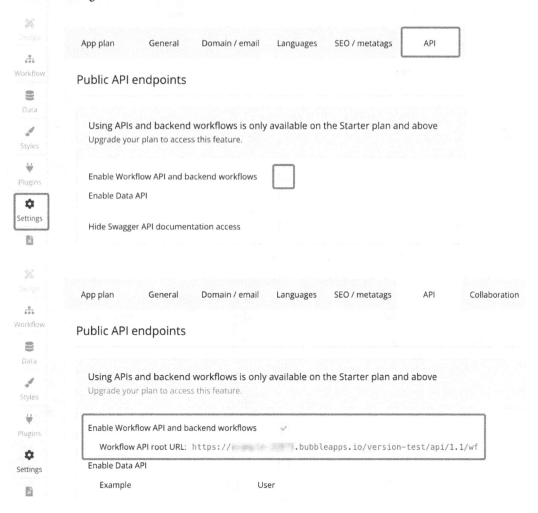

Figure 7.22: Enabling backend workflows

Once you check this option, it will expose your API URL, which can be used with external services as well to trigger backend workflows. Remember that this is a premium feature.

After activating backend workflows, to find them, you just need to click the link under your pages after the reusable elements section.

The **Backend workflows** link under the **Pages** list is shown here:

Figure 7.23: Access to backend workflows

In this area, you will be able to create new backend workflows and view all existing ones.

The **Backend workflows** area is shown here:

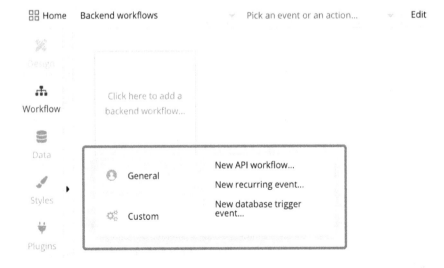

Figure 7.24: Backend workflows

As mentioned before, backend workflows don't live on the page, so they have their own location. All the workflows available will be in the same place, and you can organize them by using colors and folders. Now, let's learn about each backend workflow type.

API workflows

This feature allows you to create workflows that can be triggered by external API requests or internal schedules on your application. You can use API workflows to perform actions on your Bubble app based on parameters from other platforms or systems. For example, you can use it to create new users from a contact form filled out in another application, storing the information received from that API in your Bubble app. It is like having an endpoint of your application that can be used by other apps to trigger internal existing backend workflows.

Database events

These are backend triggers that run workflows when the database changes in a specific way. For example, you can use a database trigger event to send an email when a value on the database changes. To set up a database trigger event, you need to create it in the Backend editor and specify the trigger conditions.

Recurring events

These events can be used to run a workflow at a specific interval, for instance, daily, weekly, monthly, quarterly, or yearly. They can be considered as loops that keep going as long as you configure them to run. Another type of recurring event is a recursive workflow that can reschedule itself. Keep in mind that some of these features are only available on paid plans.

Custom events

These are events that are triggered by other workflows and can take custom parameters. You can use custom events to do the following:

- Avoid repeating work by using the same workflow in different places

- Maintain consistency and prepare for future changes by using the same workflow for similar actions

- Control the order of operations in a workflow by using the **Trigger a custom event** action

Backend workflows are very powerful but can become quite complex, so, if you are a beginner, make sure to only use it if you need it, take your time to study, play with it, and learn as you go. For now, it is just important that you know it exists, the basics of it, and how it works. I am sure you will have a lot of time to play with them in the future.

Summary

In this chapter, we've covered the basics of workflows, triggers, and actions. You learned what workflows are and how they work, how to set up workflows, and how to configure actions to run when the workflows are triggered.

You also learned how to organize workflows, edit workflows, what custom states are and how to create new ones, change existing values, and use conditionals based on custom states.

This chapter also explained how to use filters to prevent workflows from running and how to add pauses or use the results of previous steps to better control your workflows. You learned the basics about backend workflows and how they work.

Understanding these concepts will be important in helping you build multiple types of applications using Bubble and no-code, as the workflows are an essential part when it comes to building software with dynamic features, data, and logic. In the next chapter, you are going to learn how to structure and use databases, create data relationships, and a little bit about security as well. Are you ready?

8

Database Structuring, Relationships, and Security

In the previous chapter, you learned how to use **workflows** and build **logic** inside Bubble. In this chapter, you will learn about **databases** and how to structure them, create relationships, and add security rules to protect your data.

You will learn how to creatme databases and use workflows to add data to them, allowing you to store users' information. Learning how to use databases and workflows together is an essential path that will allow you to build dynamic applications inside Bubble.

In this chapter, we will cover the following:

- Bubble database overview
- Understanding data types and fields
- Building a database structure with examples
- Creating relationships
- Databases and **option sets**
- Connecting data elements within your app
- Data structure and database management
- Security best practices

Bubble database overview

Databases allow you to create dynamic applications and Bubble has a built-in database system that allows you to store information right inside your project. Not all no-code tools out there have a built-in database available inside the tool itself; sometimes, you have to connect your no-code tool to an external service that will provide the database for your no-code application, but Bubble has not only workflows and logic but also a built-in database ready for you to use. By the way, if you are wondering, you can also use external databases in Bubble if you wish to – it has both options.

What is a database?

A database is an organized collection of structured information. All the data that is stored in your application goes to a database. It is designed to efficiently manage, store, retrieve, and manipulate data, allowing users or applications to access and work with information quickly and effectively. In simpler terms, a database is like a digital shelf that stores various types of data, such as text, numbers, images, or more complex data types, in a structured manner. It enables the storage of large amounts of information and provides methods for organizing, updating, querying, and analyzing the data according to specific requirements or needs.

A database schema structure is as shown here:

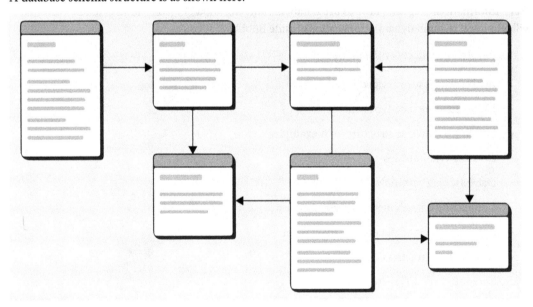

Figure 8.1: Database schema

Databases are fundamental in various applications, from basic record-keeping systems to complex enterprise solutions, forming the backbone of many software applications and websites.

The Bubble database

Now that you know what a database is and that Bubble has a built-in database, let's dive in and get familiar with this feature.

The **Data** tab main view is shown here:

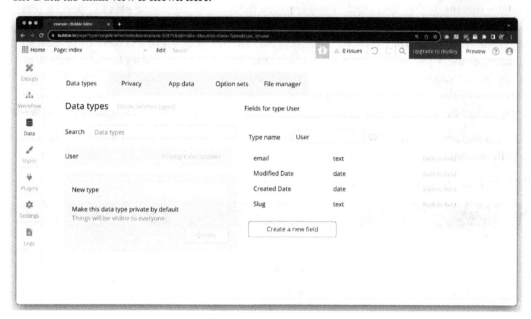

Figure 8.2: The Data tab

It is very important you get familiar with this area since most of your applications will require a database. So, the more you understand it and play with it, the better it will be for you in the future, so spend some time playing with this area while reading the book. Let's dive into each internal tab available inside the databases tab.

Data types

Once you enter the **Data** tab, the first section will be **Data types**, which is where you create and manage your database and all the fields available inside it. At the top, you can search for existing database names. This is going to be useful in the future once you have a lot of tables as it will help you locate the ones you need faster.

Inside Bubble, you only have one single database, and all the items you create are like tables. Inside each table, you can create multiple types of fields to store information, and each field can have its own type and be unique or store a list of values. By default, every new application comes with a table called "User"; this is because Bubble already configured this one for you, so you can create a login and sign-up feature. If you don't need a login and sign-up feature on your app, you can just ignore it. Notice that inside the "User" table, there are a few fields on the right side: `email`, `modified date`, `created date`, and `slug`. The `email` field is a text – it will store the users' emails when they create an account. The password is not a visible field in this case, but it is there by default, so you can store users' password information. Bubble hides it for security purposes and not even you as the app builder can see it. Now, the other fields are automatically created for every table; they store information automatically, so you can know when someone created that record or modified it. A slug is a unique identifier for that specific record – you can choose to use it or not, and it is like a unique tag.

There are also other tabs available inside the **Data** tab, which are **Privacy**, **App data**, **Option sets**, and **File manager**. These are all related to the main database; let's take a look at each one of them.

The **Privacy** tab main view inside the **Data** tab is shown here:

Figure 8.3: The Privacy tab

Privacy

The **Privacy** tab is very important. Here, we can define rules to prevent users from getting access to data they aren't supposed to.

The **App data** tab under the **Data** area inside Bubble is shown here:

Figure 8.4: App data

Mastering this area is important for your app security. In the following sections, we are going to learn how to create and set up some privacy rules.

App data

The **App data** section shows your database data in a table view, so you can see and manage every record available inside your database.

The **Option sets** tab under **Data** inside Bubble is shown here:

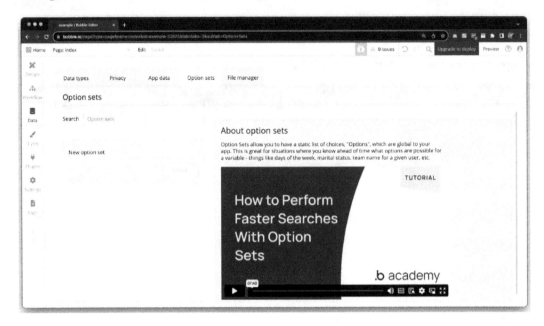

Figure 8.5: The Option sets tab

This area is very flexible. Here, you can choose how you want to visualize data, which data items to show or hide, you can create custom views, and even manage your database. For instance, you can add entries from here, simulate someone using your application, search, upload, modify, and even export data from your database using the buttons available in this area. You are going to use it very frequently when working with databases.

Option sets

Option sets are a kind of database, but they work differently. While a database is used to store dynamic and structured data, option sets are a predefined list of fixed options or choices. The main difference is this: option sets can be defined and changed by you, the app builder, but not by your users, while a database can be created and edited or even deleted by your users. So, when and why use option sets instead of a regular database? Well, an option set is usually used to add a set of constant values or choices that can be used as drop-down lists, radio buttons, checkboxes, or other selection interfaces within the application.

The **File manager** tab under the **Data** area of Bubble is shown here:

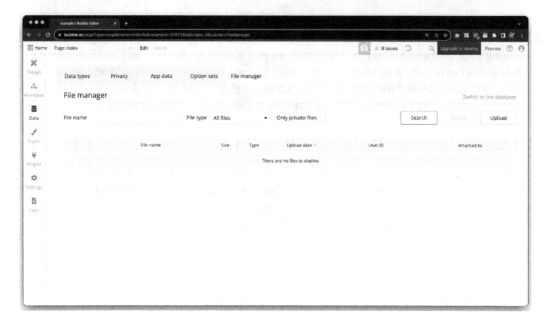

Figure 8.6: The File manager tab

The reason for using this is that sometimes it can be faster to load these options because it won't need to do a query on a database to display these options to the users.

File manager

The last but not least tab is the **File manager** tab. Here, you will visualize and manage all the files stored on your application, such as images, documents, audio/video files, and static assets. Every file stored via database or on your pages will be added here, such as visual content, user-generated uploads, media files, static resources, and exported reports or data. Inside **File manager**, you can search for specific items, upload new items, and even delete old and not used ones to save storage space. The amount of data you can store is directly related to your Bubble plan. Usually, the cheapest plan starts with 50 GB storage and it can go to 1 TB or more on a custom enterprise plan, so it is important to use your storage wisely.

Now that we did a quick tour through all the tabs inside the databases area of Bubble, you know how to navigate and go to the specific areas you need while building your no-code applications.

Creating database tables and option sets

Let's see in practice how to create new tables inside your Bubble database. You must plan your database structure before jumping into the tool, otherwise it can become a complete mess. Be careful while structuring your database, as this is a very important part of your application, and building a wrong database at first can become a nightmare in the future. So, just because it is working doesn't necessarily mean that it has been created correctly. If you are not sure, talk to a professional developer to help you with that or find tools and resources that can help you build a proper database schema.

Creating a database table

Creating a new database table is very simple – just click the **New type** field and add the name you desire, then click the **Create** button. You can also choose whether you want this table to be private, meaning only users who created the items will visualize the items they created. This is optional for now – you can always set privacy rules later as well.

An example of how to create a database type (table) is shown here:

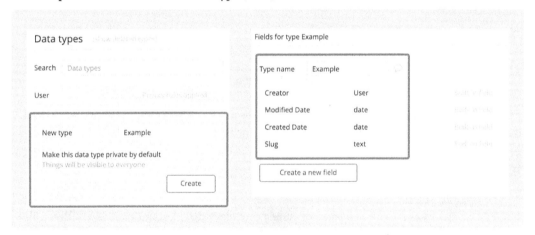

Figure 8.7: Data types

Once you click the button, you will notice that a new item will be added to the list. On the right side, you will notice that you have a new table structure with a couple of fields, such as `Creator`, `Modified Date`, `Created Date`, and `Slug`. These fields come by default and you don't have to worry about them, as you can see next to it there is text saying **Built-in Field**. Every time you see this, it means Bubble created it for you and these items come by default. Now, you can start adding fields to your newly created table. To do that, click the **Create a new field** button.

An example of how to create a new database field (column) is shown here:

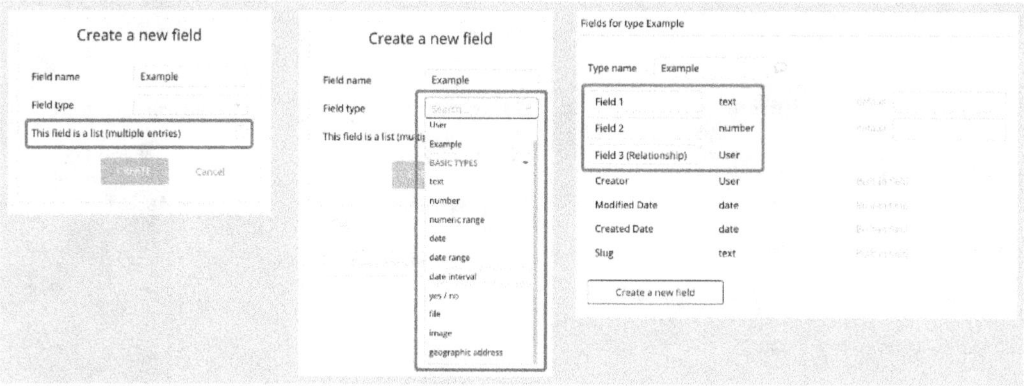

Figure 8.8: Creating a new database field

A new popup will open, and all you have to do is add a name and choose the field type. Once you click the field type dropdown, it will show you a list of options you can choose. These options define which type of information you are going to store on your database item. Choose this carefully because it will impact how your application works. The field type must be compatible with the type of data you are going to store in your database. For instance, if this field will store the username, it makes sense to choose a Text field type; if you are going to store an image, choose the Image field type; if you are going to run calculations with the values and add numbers, choose the Number field; and so on.

You will also notice other sections inside this dropdown – it all depends on your database structure. If you have option sets, you can link this new field with an existing option set, creating a relationship between this database table and an option set. This concept can also be used for other database tables, which is how you create a relationship.

A database relationship is a connection or association between different sets of data stored in a database. It allows data from different tables or collections inside a database to relate to each other. Think of a relationship as links between the data; this will enable you to reference and fetch information on your databases in a more efficient and organized way while building your application pages.

Another option you have to consider while creating a new field is whether it will be a single item or a list. If you want to store a list of information, just check the little checkbox and it will be considered as a list; if you don't check it, it will expect to receive just a single value per database entry. To create a relationship, you need to choose the field name belonging to another existing table and use it as a reference for that specific field. If you want this relationship to work both ways, you will have to create it inside each table individually. It can seem complicated, but it is not – you will get familiar with it very soon. Relationships are basically connections; it is one item referencing another item or both referencing each other. This connection between them allows you to fetch data from each item when building database queries on the frontend later on.

An example of a database relationship field is shown here:

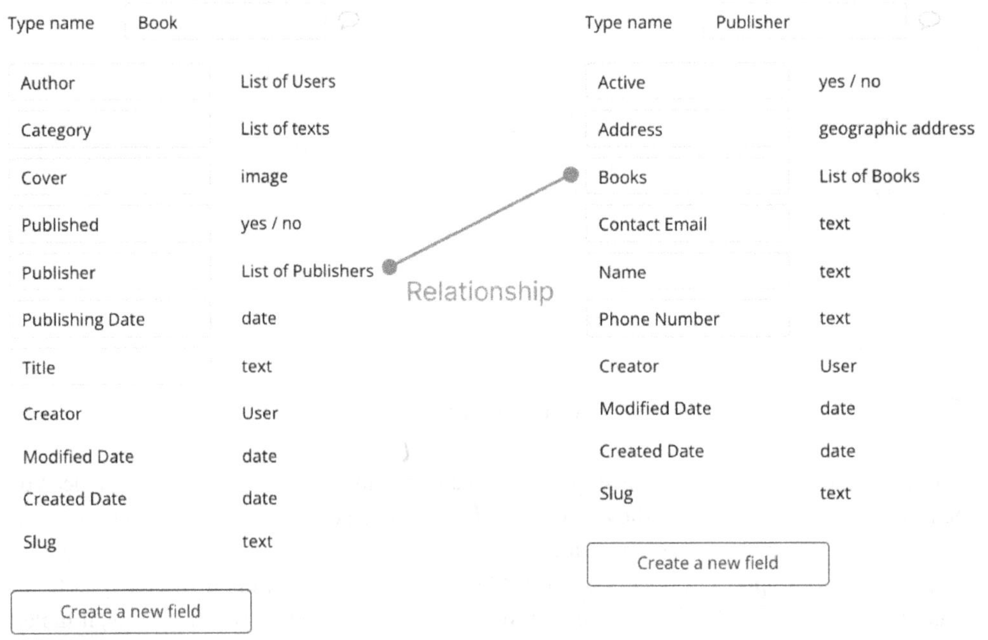

Type name	Book		Type name	Publisher	
Author	List of Users		Active	yes / no	
Category	List of texts		Address	geographic address	
Cover	image		Books	List of Books	
Published	yes / no		Contact Email	text	
Publisher	List of Publishers	Relationship	Name	text	
Publishing Date	date		Phone Number	text	
Title	text		Creator	User	
Creator	User		Modified Date	date	
Modified Date	date		Created Date	date	
Created Date	date		Slug	text	
Slug	text				

Create a new field

Create a new field

Figure 8.9: Database relationship

After clicking the **Create a new field** button, you will visualize the new fields available on the right. You can also choose to change its name if needed – just click the item name and change it; don't worry, it won't break your database. Next to the title, you will be able to see which type of item that is, according to what you chose when you created it, for instance, text, number, and so on. If you see the name of another database table next to the item title, for instance, user or any other name, it means it is a relationship field. If you see it saying **List of** and the item name, it means that it is a list type of field, which means it can receive multiple entries on the same record.

Creating option sets

When it comes to creating option sets, the process is very similar to creating database tables, but, of course, this process happens inside the **Option sets** tab. To get started, first, add the option set name and create it, then click to add a new attribute. You can create as many attributes as you want and define the attribute type by choosing one of the options listed in the dropdown.

An example of how to create option sets is shown here:

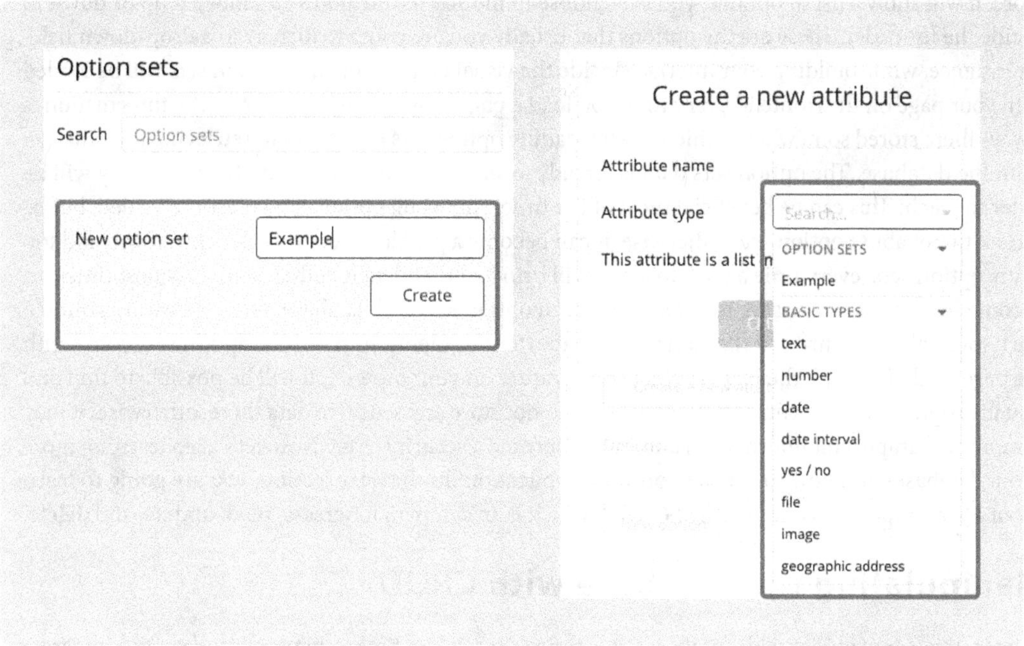

Figure 8.10: Option sets

Again, the process is basically the same as creating new tables and new items on a table. The main difference is that option sets, as the name suggests, allow you to create, options. As we can see, there is a field to add new items to that specific option set. Each option set has attributes, the ones you created before, and inside each option set, you can add value to the attributes, in case you want to pull that information when manipulating data on your frontend.

An example of how to add new option sets options as shown here:

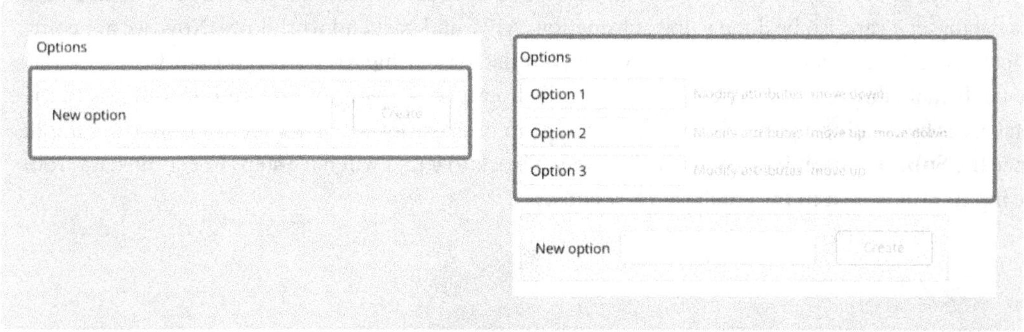

Figure 8.11: Option sets options

To create a new option, just type in the name and click **Create** – you will see it right away. If you add more, it will show a list of options; you can choose to modify its attributes and move it up or down to decide the list order. These are the options that usually you are going to display in a drop-down field, for instance, while building your interface inside the visual editor. Note that option sets will be loaded with your page on the frontend every time you load a page. It works like memory – the information is always there stored somewhere, which is why loading option sets is faster than retrieving information from the database. The option sets data is already available on the page load, therefore, they will be faster to reach. This can be beneficial most of the time, and using option sets is recommended, but be careful to not abuse option sets, otherwise, it can become a problem, too. Imagine this: if you add too many option sets, every time a page loads, it will bring all that data together, so the request time can become slower. Just make sure to balance between option sets and database items. As we are going to learn more about security in the future, it is important to mention that since option sets come with the page load, if you use the Browser Inspector feature on your browser, it will be possible to find and visualize option sets information, so make sure to not store any sensitive data there, otherwise, it may compromise important information and might become a security issue. Now, let's keep learning more about databases and how to use data on your application. In the next section, you are going to learn about a very important concept called **CRUD**, which is an acronym for **create**, **read**, **update**, and **delete**.

Manipulating the database with CRUD

Now that you are familiar with databases, it is time to teach you how to manipulate data and integrate database items with your frontend, your application pages, and interface. This practice is also known as CRUD. These operations are all available inside **Workflow Actions** and are what will allow you to build cool dynamic applications using no-code. Let's learn how to create them using Bubble.

Create

After creating the database structure, of course, you want to store data inside it, right? But how do we allow our application users to add data inside the database while using our application? We do this by simply giving them fields to type in information. The creation of data inside your database will rely mostly on your users interacting with your pages, but, of course, that is not the only way to add data to a database; it can also be done using automation, APIs, and backend workflows. Now, we are going to focus on allowing your users to add data to your database using forms. As you saw before, inputs are fields that allow users to type in things or select options. A common method to add data to the database is by creating a page where users will type in information and click on a button to submit it. Once the **Submit** button is clicked, it will trigger a workflow that will have steps to get the data from the inputs and store it in the database you choose.

An example of a form with inputs that users will use to add information to a database is shown here:

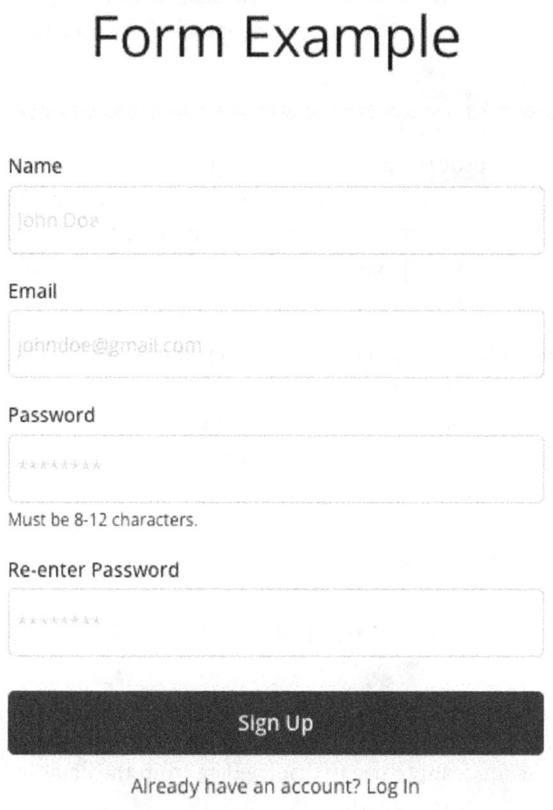

Form Example

Name

John Doe

Email

johndoe@gmail.com

Password

* * * * * * * *

Must be 8-12 characters.

Re-enter Password

* * * * * * * *

Sign Up

Already have an account? Log In

Figure 8.12: Example Sign Up form

As you can see in the image, in this scenario, the user has a few input fields to type in information; after that, it is possible to click a button to sign up. Once this button is clicked, it will trigger a workflow with steps to get each information typed in the input fields and then create a new user account using this information. This process is called **data binding** as you are connecting information provided by the user or a page to the existing fields available in the database; it is like creating an automation that will tell Bubble what to do, like if you were saying, "Hey, Bubble, get this information and store it here." As I mentioned before in previous chapters, Bubble is very powerful, but you have to tell it exactly what to do, step by step, and if you do it correctly, you will be able to build amazing things. Let's take a closer look into what these steps look like:

1. Add a new workflow to the **Sign Up** button.

2. Add a step to sign the user up.

3. Connect the input fields with the action.

4. Choose what happens next. It can be sending the user to another page or showing a success message to let the users know that the action was successfully accomplished.

Read

The **Read** action is the inverse process; instead of adding things to the database, you are going to retrieve information and display it on a page. This is also a very important thing to know since you are going to use it quite often. To add information to a page, you first need to draw your page, add the components needed to build your page layout, and the components that will receive dynamic information, such as groups, buttons, texts, and images.

Once you have your layout ready, you can start adding rules to pull information from the database. It can be a single piece of information, or it could also be a list of information – we call this a query. If you are only going to pull one piece of information, such as a name, for instance, you can use single components to do that. If you are going to bring a list of data, you can use a repeating group to do that. The repeating group can be a great option for displaying lists and you can customize it to display the amount of information you want to display on a certain page. Another component that can help with this task is the Table component. You can test which components will work best for your specific use case and layout. Be careful when bringing a lot of data to a page, as this may cause performance issues and increase the loading time, so, always plan and try to bring only the data you need when creating your queries.

Let's see in practice how to bring data from the database and display it on a page:

1. Select the component you want to use to display data from the database.

2. Click the dynamic field and start building your query.

3. Select the database type and the field you want to bring.

4. Preview your page and see whether the data is correct.

Now, let's learn how to display items on a list:

1. Use a repeating group and build a simple layout to receive data.

2. Create your query, type in the type of data you want to bring, and for that, use **Do a search for**.

3. Now, inside each component on the first element inside of the repeating group, set what type of information it will display.

4. Preview the page to see whether the data is displaying correctly.

It is very common to use filters and conditionals to shape the type of data you are going to pull from the database. Adding rules will allow you to be more specific and tell Bubble what specific information or group of data you want to display on a specific page. Bubble has several ways to filter content but remember to mostly filter data on your queries and not in a visual way. It is always best to save the work of fetching data from the database, rather than hiding it, for when after the query is done. You can also use input fields and interactions on the page and combine them with custom states to manipulate how the page works and how the data is displayed; you can use the current user information and decide to show or hide elements based on their current status, for instance, if they are logged in or not. There are many tricks and resources you can use to manipulate the page and the data. Since I can't show you everything, it is advised to invest some time in testing and discovering more about these functionalities. One good place to find more information about it is in the Bubble documentation.

Update

Once you have data on your database, it is very likely things will need to change, so, to change things, you are going to use an update operation. To do updates, the most important part is to tell Bubble what information is being edited and the new piece of information that will be replacing the old one. This time, we do a query first to locate the item and second, we bind the fields to add the new information to the database. Editing an item usually involves using input fields, but this time, the inputs will be prepopulated with the existing information, which will help us not only to display the current information to the users but also to let Bubble know what type of data we are talking about. Using inputs is not the only way to change things on the database – you could also make predefined changes just with a click of a button and in many other ways. Think about the update action as a mix of the create and the read actions. Let's take a look at a practical way you can set up this action.

1. Create the layout to display the current information to the user.

2. Define the query to display that current piece of data on the elements.

3. Create a workflow action to update the fields.

4. Set up the fields and pass on the information to the database, replacing the old data with the new data.

5. Add a final action to reset the inputs and go to another page or update the existing page with a success message.

Delete

This is one of the simplest actions because you don't have to do much to delete items; you just need to tell Bubble what data you are referring to and set up an action to just delete it. Now, it is important to know that once the data is deleted, it is going to be deleted for good, so it is a good practice to let your users know about it to prevent future headaches and mistakes in deleting important stuff. Although this action is simple, it will take a bit of extra work on the UX's part to make sure you provide a good experience to your users. Let's take a look at how to set up a delete action on your database:

1. Fetch the items from the database and display them on a page.

2. Set up a workflow to delete the items and add the delete step to it.

3. Send the user to a new page or show a success message.

It is recommended to add a pop-up window with a confirmation message before actually deleting the item. To do that, you can create a pop-up window with a message and two buttons. Instead of using a workflow action to delete the item when you click the delete button right away, you can use a different approach to first ask the person a confirmation before proceeding with the item deletion.

Once the user clicks the delete button you created next to the item that should be deleted, do not add an action on this button to actually delete the selected item. Instead create a workflow action that will trigger the pop up window you created to confirm the action. Make the delete button, via workflow, open the reveal the pop up when clicked, the next step should pass the parameter of what item is being clicked to the pop up parent group, so you have this info to use later on the next step. Now you will configure the two buttons, cancel or delete to do different things. First configure the Cancel button to close the pop up window. That will simply make the pop up become invisible again and nothing will be deleted. For the delete button then you can create a workflow action to delete the item, but then you need to refer the item you want to delete by getting this info from the parent group, which received it from the previous step. The last part is to also close the pop up, so it disappears after you click the delete button. You can go even further and create an alert to display a success message after the pop up disappear. A quick tip is also to customize the design of these buttons to make sure the delete button is different from the cancel button, the delete button usually stands out more because this is an important action and the users need to be aware of it.

If you don't want to permanently delete items for some reason, an alternative is to not use the **delete** action, but to add a database field to the data and just change its current status. So, instead of deleting an item, you are going to just say it is deleted and store this information on the item itself. When you display items on the page, you will just need to remember to use this database field to make sure you are not going to display items that were supposed to be deleted. Make sure you add filters to your query to eliminate the items that shouldn't be visible. For instance, on your database you could have a field called deleted, if it is yes, then you don't show them on your page. Note that this approach also can slow down your queries due to the amount of items being requested at the same time. This is not a common approach but it can be a helpful idea in some cases.

Privacy rules and security

Defining privacy rules has a direct relation to security in Bubble because these settings help you specify how and who exactly can view and manipulate data on your database. You may think that if a page is not showing data on a field, it might not be exposing it, but a lot of data can be loaded on a page even if it is now visible; this means that people with the right skills and knowledge could easily get access to your app's data, even if you are not directly displaying it on a page. This is why privacy rules can help, because they will prevent your data from being displayed even before the page loads, protecting it from the foundation of your database. Note that privacy rules play an important part in your app security, but that is not all – some other methods and techniques can be used to make your app more secure, some settings can be configured on the workflows, and even sometimes at the page level. If you are really worried about security or have strict compliance rules, it is recommended you search for some tools available just to help you spot vulnerable points inside your Bubble application; you can easily find some alternatives by doing a Google search.

Now, let's take a look at how you can set up a privacy rule. First, go to the **Privacy** tab; there, you will see all the tables on your database. Click on the top of the table you want to add rules to. On the right side, you will see a button to define a new rule.

The example of privacy rules is shown here:

Figure 8.13: Privacy rules

Once you click the button, just choose a name and create this new rule, and a new area on the right side will show up. In this area, you are going to see a box where you can click to add your conditional rules. You are going to create a formula that defines who can do what in that particular database table. Notice that there are two boxes – the first one is a specific rule and the second is a general **Everyone else** rule; this allows you to define specific rules and rules that are applied by the rest of the users.

An example of how to create a new privacy rule is shown below:

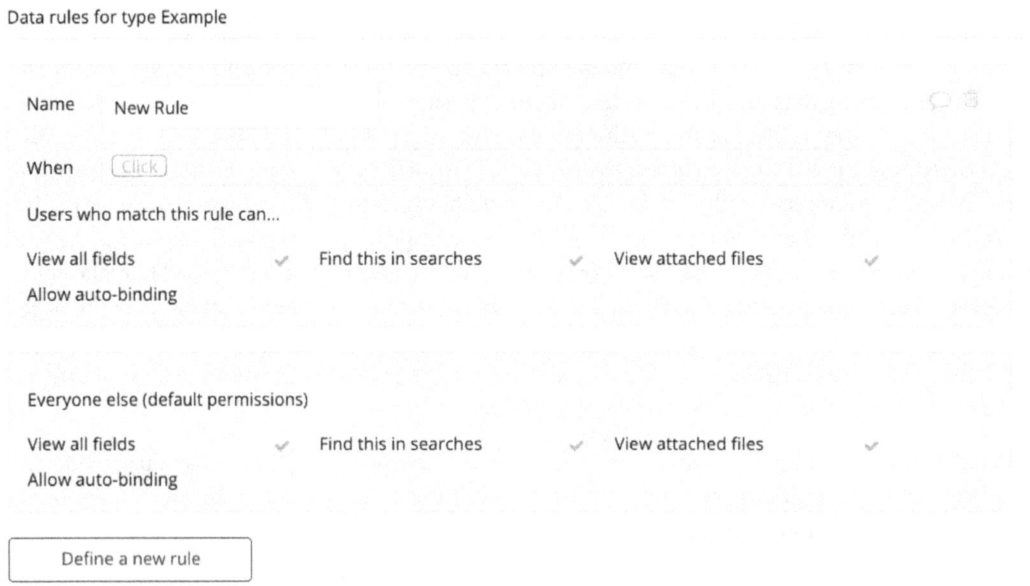

Data rules for type Example

Name	New Rule				
When	Click				

Users who match this rule can...

View all fields	✓	Find this in searches	✓	View attached files	✓
Allow auto-binding					

Everyone else (default permissions)

View all fields	✓	Find this in searches	✓	View attached files	✓
Allow auto-binding					

Define a new rule

Figure 8.14: Data rules

See, for instance, the New Rule example. Inside the **When** field, you are going to start choosing who you want to target, for instance, logged-in users, meaning only users who have an account will be able to do a specific thing. The checkbox items below are the options you can choose; for instance, you can allow users who match specific criteria to be able to view all fields, or not. To decide which options are available for that group, simply check or uncheck the options. By doing that, your users will be able to only perform specific actions while playing with your application. If you uncheck the **View all fields** option, more items will show up, meaning you can be more specific about what specific fields inside that database table can be visualized by a certain type of user.

The example of a privacy rule based on the user conditions is shown here:

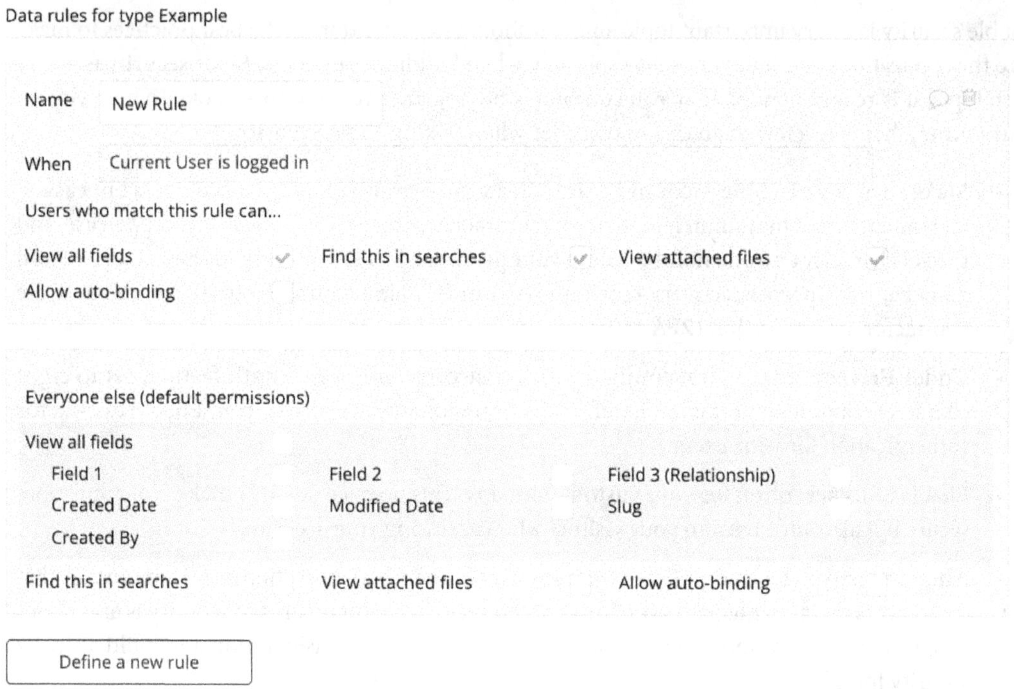

Data rules for type Example

Name   New Rule

When   Current User is logged in

Users who match this rule can...

View all fields   ✓   Find this in searches   ✓   View attached files   ✓
Allow auto-binding

Everyone else (default permissions)

View all fields

   Field 1   Field 2   Field 3 (Relationship)
   Created Date   Modified Date   Slug
   Created By

Find this in searches   View attached files   Allow auto-binding

Define a new rule

Figure 8.15: Defining privacy rules

For instance, this is an example of a rule that only allows the current users to visualize fields, find them in searches, and view attached files on that particular database if they are logged in. Everyone else won't have such permissions as all the other options are unchecked on that box. As you can see, you can create multiple rules to make it more personalized to your application needs.

Security tips

Bubble security is a very important topic and you should pay attention to the best practices to make sure the applications you build are safe, especially when building for clients. Since security is a very deep topic, it is recommended that you continue studying after reading this book. To help you on your journey, here are a few main tips to consider when talking about security:

- **Make sure your Bubble account is safe**: Create passwords that are hard to guess but easy to remember, have a minimum length of eight characters, and have one lowercase, uppercase, and special character and one number. Store your password in a safe place; you can use a password manager. Set up your recovery email under your Bubble account **Security** tab and activate two-factor authentication (**2FA settings**).

- Under **Privacy & Security**, configure the password minimum length feature, set to eight characters, a number, a capital letter, and a non-alphanumeric character. Enable two-factor authentication for your users.

- **Enable SSL encryption for your custom domains**: This practice not only makes your app more secure but also adds trust to your visitors when accessing your website.

- **Align security rules and policies with your team members and clients**: It is important that everyone is on the same page and thinks about security. Remember that security is also about people and organizations, so you also have to take that into consideration and build a strong security foundation.

- **Plan your app upfront**: This is one of the best ways to ensure security. List the important pages and types of information that need to be secured even before building your app; this will help you prevent unexpected scenarios.

- **Avoid using popups to hide important information behind them**: They can be easily removed using browser inspection tools. While this is a very common practice, it is not secure at all, so, make sure to implement other security measures to prevent users from accessing the information.

- **Avoid using plugins you don't trust or are suspicious**: Remember that every plugin added to your application is a dependency and you can't control how exactly they work, so, use plugins from reliable developers to avoid security issues and unexpected behavior on your application.

- **Make sure to use and configure your app's privacy rules**: This is a very important step and will help your Bubble app be more secure.

- **Use auto-binding when possible**: This feature automatically updates data from an input field when users leave the field (remove focus), and auto-binding fields can be controlled inside **Privacy Rules**, which can make it easier to maintain and prevent errors.

- When creating database queries, only bring the data you need to display; use filters and don't hide information just on the frontend. It can still be accessed via browser tools.

- If you are not using API features, disable **Public API Endpoints**.

- Under **Privacy & Security,** make sure to set your app to **Private App**, protect your Bubble editor, and run mode access using a username and password.

- Create server-side conditions on workflows using the **Only when** field to define rules that can limit unexpected situations or scenarios.

- **Make use of server-side redirects and rules**: Creating routes for specific pages when users try to access it will prevent users from accessing pages they shouldn't.

- **Run periodic maintenance and tests**: Under **Settings | General | App file management**, use the option to optimize the application from time to time.

- Use tools that can help you run security checks and get an automated diagnosis of what could be improved.

- **Create a deployment plan**: With a checklist, you can avoid mistakes and prevent errors, including the ones that can impact your app's security.

Making sure your application is secure is very important and is a constant process that will keep evolving as you build; take the time to learn more about it and apply the tips and advice provided in this book, as they will help you along the way.

> **Note:**
>
> If you want to dive deeper into Bubble's security I recommend you check these sources: Peter Amelie's Book - The Ultimate Guide to Bubble Security
>
> - The company Flusk
> - The company ncScale

To help you learn and understand more about databases, at the end of this chapter, we are going to share simplified versions of database structures you can use to start playing with databases inside Bubble. Note that the goal here is not to provide a full database for a full application, but just to give you an idea of how a database could look in terms of structure.

Use them just as a reference for learning purposes. In a real-world scenario, these databases would become more complex and need more columns, more fields, and more relationships if they were built and structured to hold an actual application. We intentionally removed complexity in these examples, so you do not feel overwhelmed at first. Take your time – databases can become complex and daunting, but the more you practice and get familiar with them, the easier it will become.

In the tech world, we usually call these documents database schemas, but you can use any method you prefer to structure a database; it can be a simple piece of paper, a spreadsheet, a document, or even a more sophisticated tool created to design databases. In this example, we are using simple tables to organize the database structure to make it simpler to understand.

The first example is an application similar to Instagram, but again, very simplified to the core features. Let's look at the database structure.

The example database structure is shown here:

Simple App similar to Instagram

User (default)	
Name	Text
Email	Text (Built-in field)
Bio	Text
Handle	Text
Profile_Picture	Image
Followers	List of Users (Relationship)
Following	List of Users (Relationship)
Posts	List of Posts (Relationship)
Created Date	Date (Built-in field)
Modified Date	Date (Built-in field)
Slug	Text (Built-in field)

Post	
Post_Content	Post_Content (Relationship)
Comments	List of Comments (Relationship)
Likes	List of Users (Relationship)
Creator	User (Relationship) (Built-in field)
Created Date	Date (Built-in field)
Modified Date	Date (Built-in field)
Slug	Text (Built-in field)

Post_Content	
Content_Description	Text
Content_Image	Image
Creator	User (Relationship) (Built-in field)
Created Date	Date (Built-in field)
Modified Date	Date (Built-in field)
Slug	Text (Built-in field)

Comment	
Content_Info	Text
Post	Post (Relationship)
Creator	User (Relationship) (Built-in field)
Created Date	Date (Built-in field)
Modified Date	Date (Built-in field)
Slug	Text (Built-in field)

Figure 8.16: Database structure for a simple app similar to Instagram

The first database column is User, which is already available inside Bubble when you build a new application, so for this one, you will only need to create the extra fields matching each of the fields listed in this table.

A quick piece of advice is to start first by creating all the columns, and then populating the internal fields after. Since there are some relationship fields, it is best to have all the tables created first so you can find them and reference one another while creating new items with relationships. In a practical example, create the database collection items for Posts, then Post_Content, Comment, and so on. Later, go inside each item and add the items as the image here suggests. Another thing to pay attention to is when there are lists, make sure to check the **This field is a list (multiple entries)** checkbox, so that the item can receive multiple entries and be an actual list. Now, head to your Bubble account and start creating this database structure. It will be fun!

Here is an example of how that database structure will look when applied inside Bubble:

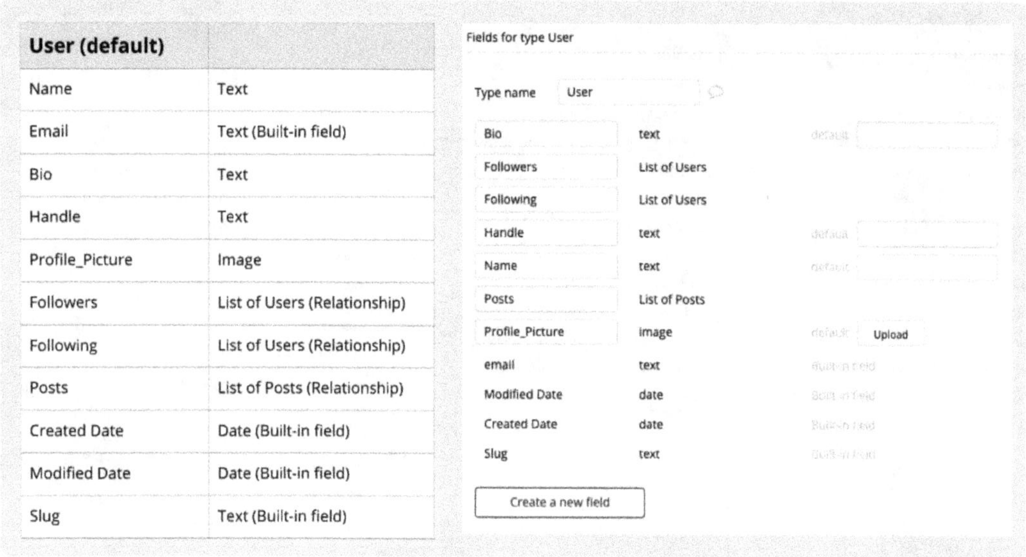

User (default)	
Name	Text
Email	Text (Built-in field)
Bio	Text
Handle	Text
Profile_Picture	Image
Followers	List of Users (Relationship)
Following	List of Users (Relationship)
Posts	List of Posts (Relationship)
Created Date	Date (Built-in field)
Modified Date	Date (Built-in field)
Slug	Text (Built-in field)

Fields for type User

Type name   User

Bio	text	default
Followers	List of Users	
Following	List of Users	
Handle	text	default
Name	text	default
Posts	List of Posts	
Profile_Picture	image	default Upload
email	text	Built-in field
Modified Date	date	Built-in field
Created Date	date	Built-in field
Slug	text	Built-in field

Create a new field

Figure 8.17: Database structure inside Bubble

Now, just continue this process until you have the entire database ready! After that, you might also want to test your other skills acquired in this book and connect the database with the frontend, creating logic and workflows to make the database receive data. You might also want to play with security rules, maybe add a few option sets, and make things a little more interesting. Keep learning and practicing.

Before we finish this chapter, I also want to share another database structure, just in case you want to continue learning and playing even more with Bubble. This time, this structure is a very simplified version of LinkedIn, of course, with the bare minimum of it, but enough for you to learn a bit more. Let's take a look at it:

The example database structure is shown here:

Simple App similar to LinkedIn

User (default)		Experience		Post	
First Name	Text	Company_Name	Text	Post_Content	Text
Last Name	Text	Company_Logo	Image	Post_Image	Image
Email	Text (Built-in field)	Creator	User (Relationship) (Built-in field)	Comments	List of Comments (Relationship)
Profile_Picture	Image	Current_Position?	Yes/No	Post_Likes	List of Users (Relationship)
Bio	Text	Job_Title	Text	Creator	User (Relationship) (Built-in field)
Headline	Text	Job_Description	Text	Created Date	Date (Built-in field)
Handle	Text	Location	Geographic Address	Modified Date	Date (Built-in field)
Experiences	List of Experiences (Relationship)	Start_Date	Date	Slug	Text (Built-in field)
Invitations	List of Invitations (Relationship)	End_Date	Date		
Posts	List of Posts (Relationship)	Modified Date	Date (Built-in field)	**Comment**	
Created Date	Date (Built-in field)	Created Date	Date (Built-in field)	Comment_Content	Text
Modified Date	Date (Built-in field)	Slug	Text (Built-in field)	Post	Post (Relationship)
Slug	Text (Built-in field)			Comment_Likes	List of Users (Relationship)
				Creator	User (Relationship) (Built-in field)
				Created Date	Date (Built-in field)
				Modified Date	Date (Built-in field)
				Slug	Text (Built-in field)

Figure 8.18: Database structure for a simple app similar to LinkedIn

Just follow the same process with this example as well. The only changes are the fields and the structure, but the concept is the same. Try creating forms and fields that can connect with the same database items and, by using logic and workflows, add data to the database as you play with your application.

If you wish to keep learning more about databases, first continue practicing, and second, head to the official documentation for more in-depth information and database examples, which is available at this link: https://manual.bubble.io/help-guides/getting-started/building-your-first-app/database-structure.

Summary

In this chapter, we did a quick overview of databases and the main topics you need to know to get started, and you learned what a database is, how databases work, and how to create your databases inside Bubble.

You learned about database structures, database schemas, and the difference between a database and option sets, as well as how they work.

You also learned what relationships are and how to create them.

This chapter also explained what CRUD is and how to integrate database items with pages, user workflows, and logic to manipulate data from your database, creating dynamic applications inside Bubble. You also learned about privacy rules, what they are, how they work, and how to set up security rules to ensure you build projects that are not only functional but also secure.

Understanding these concepts will be important for creating dynamic projects that require a database. In the next chapter, you are going to learn how to work with plugins and APIs.

9

Extending Functionality with Plugins and APIs

In this chapter, you are going to learn about **plugins** and **application programming interfaces** (APIs). You'll gain hands-on experience in navigating Bubble.io's extensive plugin ecosystem, how to leverage external functionalities, and connecting your applications with external services using APIs. By the end of this chapter, you'll be equipped with the tools and knowledge to build web applications utilizing APIs and plugins. You'll be able to broaden the horizons of what your applications can accomplish and unlock a multitude of possibilities to amplify the capabilities of your web applications with these amazing resources.

In this chapter, we're going to cover the following main topics:

- Exploring the Bubble.io plugin ecosystem
- Installing and configuring plugins
- A curated list of popular and useful plugins
- Connecting Bubble.io with popular APIs and services
- Enabling data synchronization and communication
- Enhancing app functionality through integrations
- Exploring popular APIs and finding suitable ones

Exploring the Bubble.io plugin ecosystem

The Bubble.io plugin marketplace offers a diverse variety of plugins and resources you can use to create even more powerful types of applications. Using plugins can be a great way to add extra functionality to your application and speed up your app development.

There are a couple of ways you can find and add plugins to your existing application. Let's take a look at the first option:

- Under the **Design** tab, go to the UI components sidebar and locate the **Install More** button. Click the button and a new window will open up showing a list of plugins.

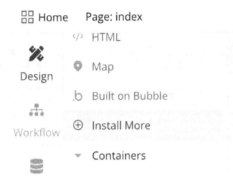

Figure 9.1: The Install More button

Once you install new plugins, if these plugins are UI components, their names will show up in the UI components sidebar.

The second option involves doing the following:

- Head to the **Plugins** tab on the left, and, once you are on the **Plugins** page, click the button to install plugins and a list of plugins will open up.

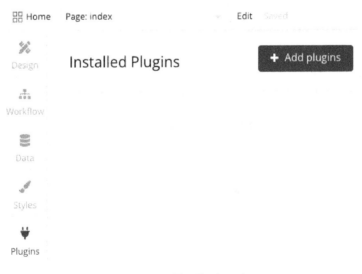

Figure 9.2: The Plugins tab

Options 1 and 2 are inside your Bubble editor and the most commonly used ways to find and add plugins. Once you click either the **Install More** button or the **+ Add Plugins** button, it will open a pop-up window that allows you to search, filter, and install new plugins to your application.

The **Install New Plugins** pop-up window is shown here:

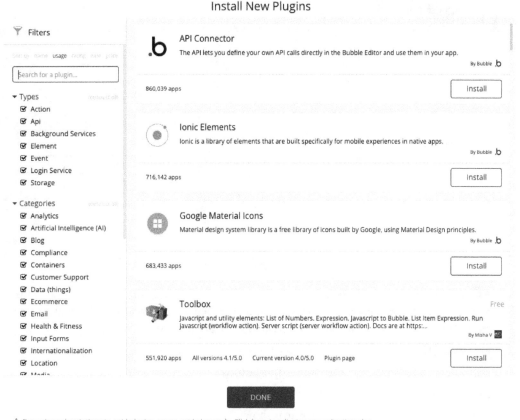

Figure 9.3: Install New Plugins

On the left, you will find filters; you can sort plugins by **Name**, **Usage**, **Rating**, **Date**, and **Price**. Yes, not all plugins are free; some are paid, and sometimes plugins are paid just once to be used while others are paid monthly. You will see the price at the top of each plugin card – it will indicate whether it is free or paid and how much it costs. Below the sorting filters, you will find a search bar where you can type in keywords to help you find plugins. After the search bar, you can use a few checkboxes to help you filter plugins by **Type**, **Categories**, **Price,** and **Built by**. Remember that there are plugins created officially by the Bubble team and others created by individuals, developers from agencies, and the community.

The following is an example of two plugins, one free, created by Bubble, and the second a paid one created by a no-code agency called Zeroqode.

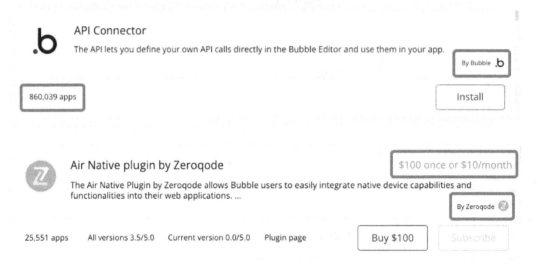

Figure 9.4: Plugins example

As you can see, there is a little indicator on the side of each plugin showing who created the plugin, which is always good to know before you install a plugin; make sure you install plugins created by developers you are familiar with and know are reliable.

Besides the two ways you already learned, there are a few more ways to find plugins:

Find plugins on the Bubble website via the marketplace. Access `https://bubble.io/plugins` and look at the available plugins created by the official Bubble team or by the community.

The Bubble **plugin marketplace** website is shown here:

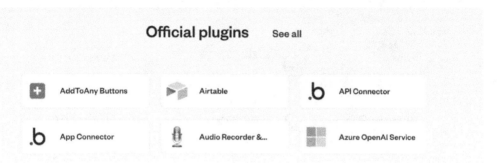

Figure 9.5: The Bubble plugin marketplace

While navigating the plugins list on the Bubble website, you can find a list of all the official plugins developed by the Bubble team; these are reliable plugins to use since they are made by Bubble. If you click each of the plugins, it will redirect you to the documentation page, which can teach you more about how the plugin works and how to use it. If you like it and want to use it, use the first or second method explained previously and type the plugin name on the search bar to locate and install it on your Bubble application.

The Bubble team has developed a lot of nice plugins, but you will find way more plugins available that were created by the community (developers, agencies, and individuals). It is awesome that Bubble has such a great community, as this has allowed more people to contribute and collaborate with the tool and expand what types of apps you can build. On the Bubble Plugins website, you can also see a list of all the plugins created by the community and find good ones to use.

The Bubble plugin marketplace with community-created plugins is shown here:

Figure 9.6: Install new plugins

On the website, you can use a search filter and a category filter to find plugins. If you click an item, it will send you to the plugin's official page, where you can find more information about the plugin and also find instructions on how to use it, the author name, a link to their profile, and sometimes their contacts or website, how many people have installed the plugin, when it was published, comments from the community, and ratings.

The Bubble plugins page is shown here:

Figure 9.7: Plugins page

From this page, you can also click a button to install this plugin to one of your applications; just select the application name in the drop-down menu and then click to open the editor. You can also see the plugin source code and sometimes even fork it, meaning you can clone the plugin and create your own version based on the source code available. Not all plugin developers will allow you to fork the plugin for obvious reasons, especially if it is a paid one, but it is nice to know that some will allow it. You can also choose to create your own plugins from scratch if you wish to, but that is a topic for another book.

Installing and configuring plugins

Now that you know a couple of different ways to find plugins, let's get hands-on experience as we walk through the step-by-step process of installing and configuring plugins.

Once you find a plugin you want to use, installing it is super easy; if it is free, just click the button to install it. If it is paid, just go through the checkout process and follow the steps.

The example of a plugin under the **Install New Plugins** list is shown here:

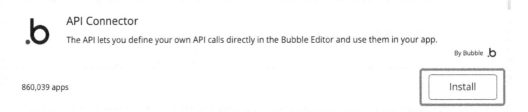

Figure 9.8: Install New Plugins

Once the plugin is installed, it will be available under the **Plugins** page; just click the **Plugins** tab to go to the **Plugins** page.

Here is an example of the **Installed Plugins** page under the **Plugins** tab:

Figure 9.9: The Plugins page

On the left sidebar, you will see the page with a list of all the plugins you have installed on your Bubble application. On the right side, you will see more information about each plugin. If you have more than one installed, just click the plugin name on the left sidebar and it will change the information to that specific plugin.

Under the plugin information area, you will see the plugin title, who created it, a little description of the plugin, and so on. Each plugin can be slightly different depending on how the developer created it. Some will have more information and even instructions or links to find instructions and others won't. There is usually a link at the top right to find more information about the plugin; the link is sometimes called **Reference**, sometimes called **Plugin** page; you can click here to learn more about the plugin.

To uninstall a plugin, you just have to click the **Uninstall** button, which is located on the right side of the page. It is very simple. The plugin will be removed. To add more plugins, you simply can click the **+ Add plugins** button again to find new plugins and keep adding more to your app. Sometimes, you will also see settings and extra fields under the plugins, which means you have options to configure your plugin – this will vary from plugin to plugin. You will also see a place to leave comments and ratings about a specific plugin, which you can use to help other Bubblers avoid bad plugins and to thank the author for creating good ones that helped you build your app.

Plugins are a dependency; when you install a plugin, your application will rely on it, so make sure to keep track of your plugins and whether there are updates. A good plugin receives updates when needed, and bad plugins (except when not needed) are old and don't receive updates; avoid these and avoid plugins from suspicious sources or unknown or unreliable developers. It is always best to use plugins that are official, well rated, developed by famous developers and agencies, constantly updated, and being used by a great number of people; these are all indicators that can help you find and select good plugins. It is important to make sure your plugins are working by testing and checking them from time to time and seeing whether there is an update; if there is, plan the maintenance to update it and test it and see whether anything changed that could break your application; this is going to be a constant part of your application maintenance.

Once your plugins are installed, you will have to either configure them or just simply use them and add them to your page. Some plugins run in the background, some will need to be applied to the page but won't be visible to the users (only you will see them as the application developer), and some others are components that you are going to use on your layouts and users will click and interact with. When you install plugins, some will show up on the UI components list for you to manipulate them.

The UI components added by the plugins installed on your application are shown in the following:

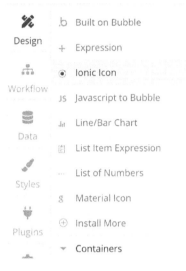

Figure 9.10: Example of UI components from plugins

Plugins that are UI components, such as **Ionic Icon**, are new components that can be used and dragged to the page. Once you add them to the page, you will see their property editor, where you can configure them according to what they offer.

The design canvas with a UI component from a plugin (**Ionic Icons**) added to the page is shown here:

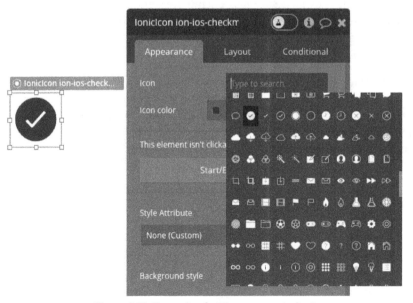

Figure 9.11: Example of a UI component plugin

When plugins have settings to be configured, these settings will be visible under the **Plugins** tab once you click on the plugin name at the left sidebar. Sometimes, you have to click buttons to expose these settings, and sometimes there will be fields you have to configure by providing extra information in order for the plugin to work.

An example of a plugin that required extra configuration is shown here:

Figure 9.12: Stripe's plugin settings

The most reliable source of information about how to use and configure a plugin is the **Plugin** page or the developer's website. If in doubt about how to use a certain plugin, refer to these places to find information or post on the community page to get help.

Now that you know the ins and outs about how to find, install, and configure plugins, let's take a look at the most popular and useful ones.

A curated list of popular and useful plugins

As you may have already noticed, there are thousands of plugins available inside the Bubble plugin marketplace, which is a good thing, but at the same time, it may make it harder to find and select the best plugins.

To help you on the journey of discovering good and useful plugins, here is a list to help you get started:

- Plugins to help with general tasks:
  - Top 15 Plugins for Bubble – Bundle by Zeroqode
  - Toolbox by Misha V

- Plugins to connect with AI services:
  - Openai & ChatGPT by Zeroqode
  - OpenAI Assistant API by Pilot & Launch
  - OpenAI DALL-E – Generate AI images by Pilot & Launch

- Plugins to track user access and behavior:
  - Google Analytics by Bubble
  - Hotjar by Zeroqode

- Plugins to add new useful UI components:
  - Air Date/Time Dropper by Zeroqode
  - Air Date/Time Picker by Zeroqode
  - Multiselect Dropdown by Bubble
  - Ionic Elements by Bubble
  - Google Material Icons by Bubble
  - Feather Icons by AirDev
  - Lottie by Goodspeed

- Plugins to add search features:
  - Search & Autocorrect by Zeroqode
  - Fuzzy search & Autocomplete by Zeroqode
  - Algolia Search V2.0 by Zeroqode

- Plugins to accept payments:
  - Stripe by Bubble
  - Paypal Checkout by Copilot
  - Razorpay by Team Visual

As you may have already noticed, there are thousands of plugins available on the Bubble plugin marketplace; this list is just a small selection. If you want to continue finding useful and cool plugins, the best way to do so is to search, install, and test new ones. Here are a few plugin creators you might like to visit to discover more plugins: Zeroqode, AirDev, Copilot, TechBlocks, EazyCode, and Pilot & Launch. Happy hunting!

Connecting Bubble.io with popular APIs and services

Using APIs allows you to integrate Bubble with external services, which will open new possibilities when building applications using no-code.

What is an API?

An API is a way for computer programs to talk to each other. Picture it like a bridge, a connection between two applications. It is like two applications using walkie-talkies and talking to each other from time to time. APIs are used in almost all software, websites, mobile apps, and computer games. You can use API integrations to show data stored in another platform on your Bubble application page, for instance, or use it to send data from your Bubble application to a third-party source.

Another example is, if you want to build a website that shows the weather forecast, you can use an API that fetches the weather data from a third-party source and displays it on your website. In summary, API integrations allow you to search for some data, read information, and create, modify, and delete something on a third-party service or from a third-party service on your application via a RESTful interface. If you want to dive deeper into this subject, search for the term "REST APIs."

Here are some examples of what you can do with APIs:

- **Integrating with payment gateways**: You can use the **API Connector** plugin to integrate your Bubble app with payment gateways such as Stripe or PayPal. This allows you to accept payments directly from your app.

- **Integrating with social media platforms**: You can use API Connector to integrate your Bubble app with social media platforms such as Facebook or Twitter. This allows you to post updates, retrieve data, and interact with users on these platforms.

- **Integrating with email marketing services**: You can use API Connector to integrate your Bubble app with email marketing services such as Mailchimp or SendGrid. This allows you to send automated emails, newsletters, and other marketing campaigns directly from your app.

- **Integrating with e-commerce platforms**: You can use API Connector to integrate your Bubble app with e-commerce platforms such as Shopify or WooCommerce. This allows you to manage your online store, track orders, and process payments directly from your app.

- **Integrating with other web services**: You can use API Connector to integrate your Bubble app with other web services such as Google Maps, Twilio, or Slack. This allows you to add new features and functionality to your app.

Note that some APIs and services online can be free and others will be paid; some services will provide an API for free but you pay for the tool you use, while other services are actually built to be APIs and you will pay by usage or requests. When considering APIs, extra services, and integrations, always be aware that new costs may apply and it will be part of your app costs and operation costs.

As you can see, using APIs can be very powerful and useful while building applications with no-code. Now, let's dive into how to actually configure and use APIs.

Enabling data synchronization and communication

The API usage inside Bubble can happen in multiple ways. The most common way is by using the API Connector plugin, which was developed by the Bubble team to allow you to create and configure custom API connections between your application and third-party services.

The API Connector plugin is shown here:

Figure 9.13: The API Connector plugin

Imagine this plugin as a blank canvas – this plugin will allow you to create and configure API connections from scratch with any API you want to use. This setup will be discussed later.

The second most common way of adding API integrations to your Bubble application is by finding specific plugins that were built with this in mind. These plugins were created by developers to specifically help you connect with a specific tool or service. For instance, for payment integrations, there is a plugin called Stripe, which means that this plugin was designed to help you integrate your app with the payment service from Stripe. It is still an API integration plugin, but, in this case, this plugin only integrates with Stripe.

API plugins are shown here:

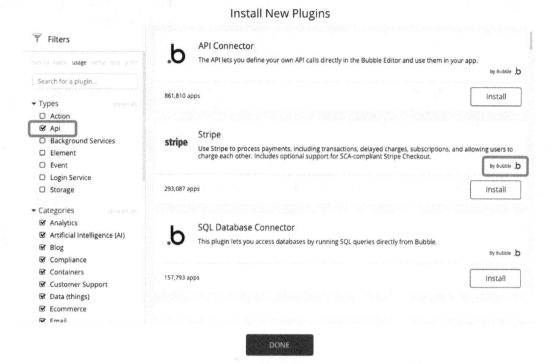

Figure 9.14: Plugin marketplace – API plugins

> **Tip**
> When searching for plugins on the plugin marketplace, choose the **Api** filter type and it will expose only plugins that were created to serve as API integrations.

Sometimes, these plugins are created and maintained by the companies that provide a service online that has API integration, but nowadays, most **Software as a service (SaaS)** companies and tools online will provide an API of some sort. Other times, these plugins are created by agencies or developers in the community, which means that these plugins are not official; they can be good and work, but they weren't created by the company providing that service, so it is good to be aware of that. It is also important to say that there are other ways to use and integrate APIs into your app, but knowing these two ways will cover most of the scenarios and use cases. Another example is to create your own plugin.

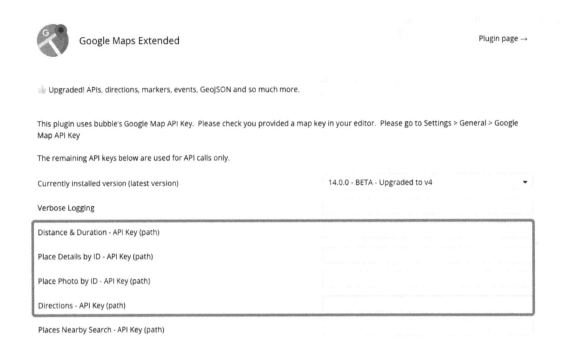

Figure 9.15: Plugin settings

When you install a plugin created to help you integrate with an API, most of the time, you will have to configure a few settings inside the **Plugin** page. Usually, you will be asked to provide your API keys – these keys are the way APIs integrate; they're like a username and password that will be used between Bubble and the tool you want to connect to. The API keys are usually available inside your account on the service you want to use; for instance, if you want to integrate Bubble with Google Maps, your API keys will be available and generated on your Google Developer account, you will have to generate API keys or find them from your Google account, copy that information and paste inside the plugin settings on Bubble. Note that you should be careful with your API keys, store them in a safe place, and do not share them with anyone. Keep your keys safe, otherwise, others might be able to use them to access your data or consume your API credits.

Now, you must be wondering why there are different ways of adding APIs to your Bubble application. As mentioned before, the most common way to connect with an API service is by adding the API Connector plugin, but APIs in general can be a little tricky to understand and set up, especially if you are not familiar with it. When using API Connector, you may have to understand exactly how the API works and how to create a request from scratch, which can be difficult for some. When using plugins that come with the name of the service, part of that configuration work was previously done by the plugin developer, which means that the integrations available on that particular service are more ready to be used and user friendly. Another great way to use APIs and build automation is by using tools such as Zapier and Make – these tools can make the integration simpler and be an alternative for you, but just bear in mind that this may bring additional costs to the project. Search for the name of these two tools in the plugin marketplace and give it a try.

In summary, using the plugins that were made to help you integrate into a specific service is usually easier to set up and use, but for some, it can be more limiting, which is why you can do it multiple ways.

Enhancing app functionality through integrations

Now that you understand how to find API plugins, how they work, and the different methods of using an API, let's break it down into practical steps and show you how to interact with the API Connector plugin.

As mentioned before, you must first install the API Connector. Once it is installed, you will head to the **Plugins** tab and locate this plugin; click the plugin page to expose its settings.

To add a new API call, you have to click the **Add another API** button.

Figure 9.16: Add another API

This will create a new box where you can set up your API connection as shown here:

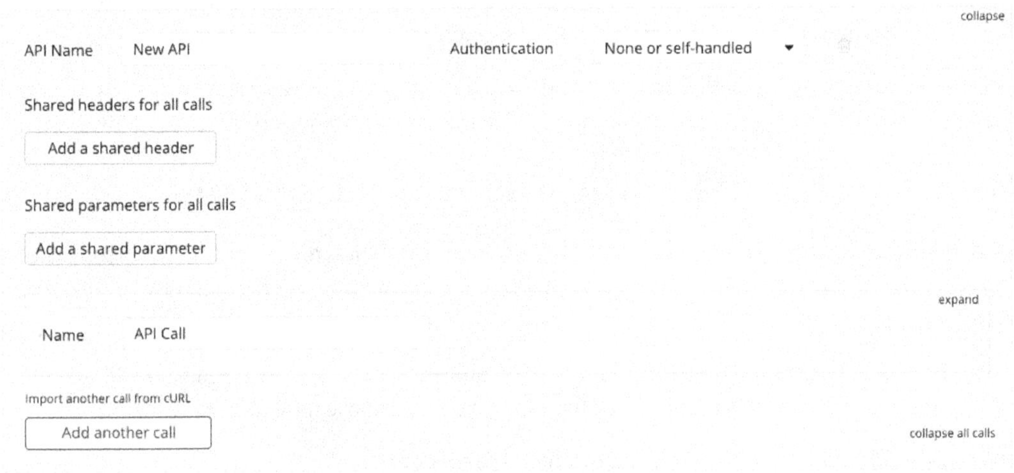

collapse

API Name   New API   Authentication   None or self-handled   ▾

Shared headers for all calls

    Add a shared header

Shared parameters for all calls

    Add a shared parameter

expand

Name   API Call

Import another call from cURL

    Add another call   collapse all calls

Figure 9.17: API settings

The fields are explained here:

- **API Name**: The first field is **API Name**, which can be used to identify your API connection later on. Add a name in this field that will help you remember what this API does and what service you are connecting to. This name will appear on other areas of Bubble while building your app, so make it recognizable.

- **Authentication**: This is the method used by this API to accept the integration, such as a user or password. Some APIs will require authentication, while others don't. Here are a few examples: **OAuth2 Custom Token**, **None or self-handled**, **Private key in URL**, **HTTP Basic Auth**, and so on. Whenever you choose a different type of authentication, new fields will show up asking you to fill out the required information. How do you know which one to use? Usually by looking at the API documentation. It might be helpful to also use tools such as Postman to help you with API setup. If there is no documentation, community, support, and friends are your best options to find out.

- **Shared headers and shared parameters**: They are reusable data (HTTP headers and parameters) that are sent with every API call and can be used to provide additional information about the request or the client making the request. For example, shared headers can be used to specify the content type of the request, the authentication token, or the user agent. Shared parameters can be used to provide additional data that is required by the API, for example, to specify the API key, the page number, or the search query. To use shared headers or shared parameters in the Bubble API Connector plugin, you need to define them once and then reference them in your API calls. When you reference a shared header or shared parameter in an API call, Bubble automatically includes it in the request. These fields are optional and will vary depending on the type of API you are using.

API calls

Inside each API block that you add to your API Connector, there will be, by default, one call. To view its settings, you need to click on the right link to expand it. To add a new one (you can have multiple), click the **Add another call** button. Since calls will show more settings, you can choose to expand or collapse them. The same happens with each API block you configure.

An example of an API call is shown here:

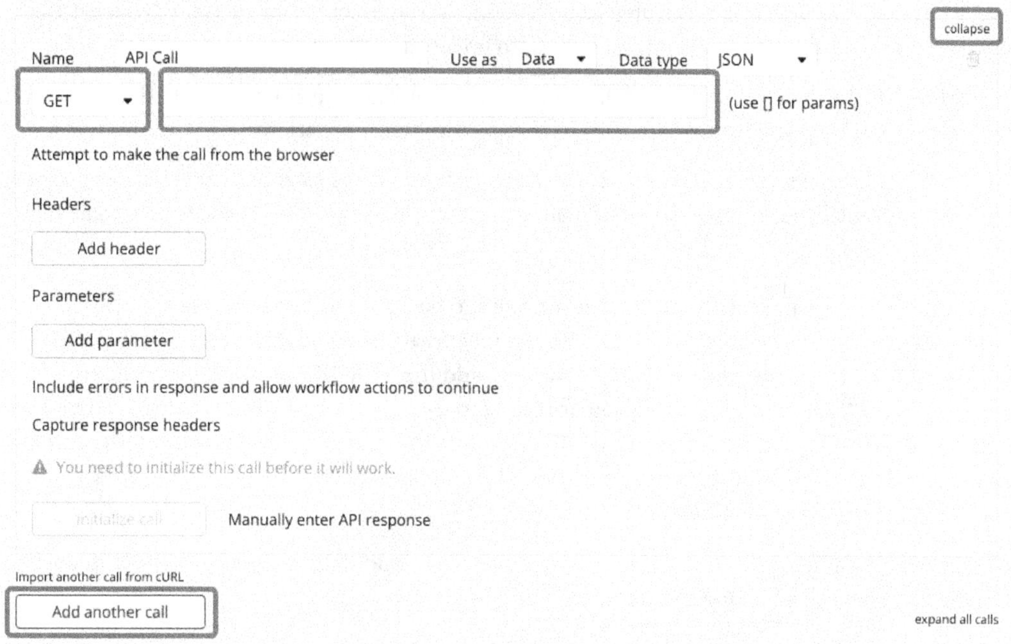

Figure 9.18: API Call

Once you add a new API call, you will need to configure each setting in order to be able to use it. The first part will be to define the **API Call** name, and then you will need to choose how you are going to use it. For this, there are two options: **Data** or **Action**. If you want to use the API information inside your pages, use the **Data** option, it will appear in the **Get data from an external API** drop-down menu while working with components. If you want to use the API data on your workflows, use the **Action** option, which will appear under the **Plugins** section of the **Actions** drop-down menu inside **Workflow**.

Next, you need to define the **Data type** field; in most cases, it will be **JSON**, but in case it is not, you can choose other options such as **XML**, **Image**, **Number**, **Text**, **File**, or **Empty**.

While using APIs, there are several methods you can use depending on what type of operation you want to perform. Here is a list of all the options available and what they can be used for:

- **GET**: This is used to retrieve data. For example, if you want to get information about a user from a database, you would use a GET request to retrieve that data.

- **POST**: This is used to create new data. For example, if you want to add a new record to a database, you would use a POST request to create it.

- **PUT**: This is used to update existing data. For example, if you want to update a user's information in a database, you would use a PUT request to update it

- **DELETE**: This is used to delete data from an API. For example, if you want to delete a user from a database, you would use a DELETE request to remove that user.

- **PATCH**: This is used to update a part of an existing resource. For example, if you want to update only the email address of a user in a database, you would use a PATCH request to update that specific field.

These are all HTTP request methods, in case you want to be a little more technical. As you can see, they are basically similar to what you would expect in a **create, read, update, and delete (CRUD)** operation. After configuring the method you have to add the API URL, this will be the address to establish a connection with the service you are going to use. This information is usually provided on the API documentation from the service you are going to connect to (e.g., Google, Stripe, etc.). It looks something like this: `https://api.twitter.com` or `https://api.hubapi.com`. These are also known as endpoints.

You might also want to configure headers and parameters – to do that, click the button to add new fields; they will require a key and a value to be filled out.

Headers contain information about the client and the request itself. You can check them to be private, which prevents this data from being sent to the client side on the browser.

Parameters are extra variable elements on your API query; for example, you can use parameters to filter data, sort results, or limit the number of results returned from that specific call. If the **Optional** checkbox is selected, the parameter is not required to make the API call and becomes a placeholder; if the checkbox is not selected, the parameter is required to make the API call and used as the default value. **Allow blank** if checked, will prevent a parameter default value used for initialization from being sent to the client or workflow inputs.

There are a few extra options you can check/uncheck if needed:

- **Include errors in response and allow workflow actions to continue**: You can use this option if you are a more advanced user and want to check errors

- **Capture response headers**: This is also a more advanced option that will bring more data from the parameters during the API call, in case you need them

Once your API call is configured, you need to first initialize the call to test it and see if everything is working properly. Click the **Initialize Call** button. A new window will open showing the result. If there is anything wrong, you will get an error – read it to try to figure out what is wrong. You might need to read the documentation if you get stuck.

If everything goes well, it will show a modal window with the request results as shown in the following:

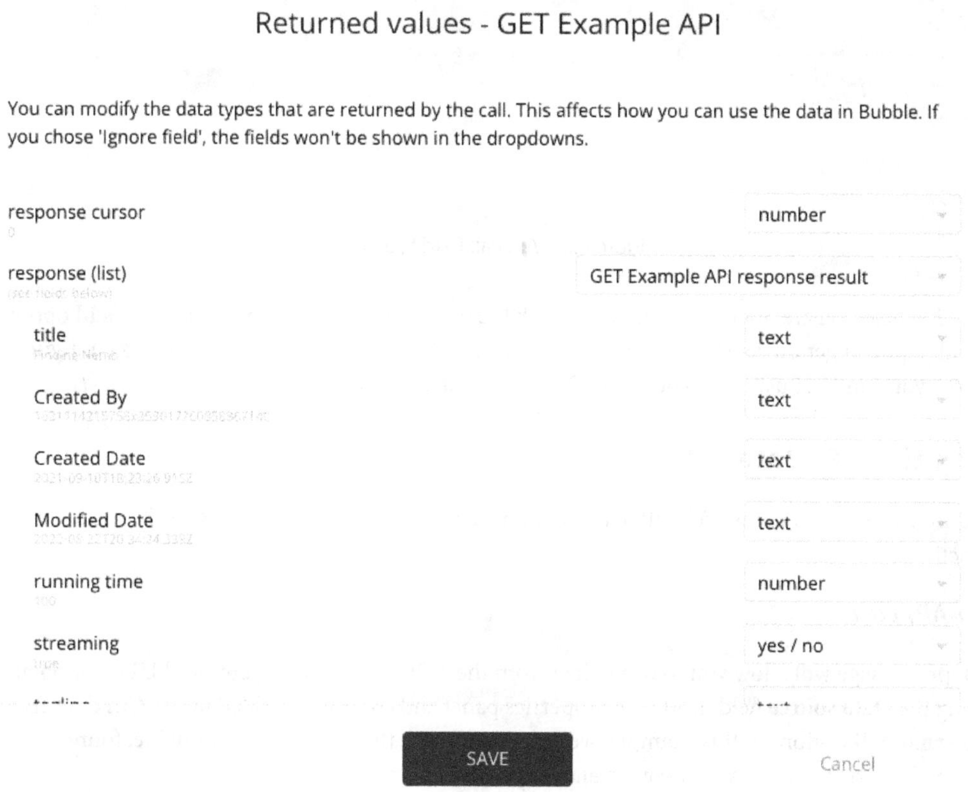

Returned values - GET Example API

You can modify the data types that are returned by the call. This affects how you can use the data in Bubble. If you chose 'Ignore field', the fields won't be shown in the dropdowns.

response cursor	number
response (list)	GET Example API response result
title	text
Created By	text
Created Date	text
Modified Date	text
running time	number
streaming	yes / no

SAVE   Cancel

Figure 9.19: The API call – Returned values

Here, you can see the returned fields and check the ones you want to use on your Bubble application. Check the fields and their types – you can click the drop-down button to change, for instance, **text** to **date**, **number**, or **image**; you must make sure the fields match the type of data you are receiving correctly.

The API request results and fields are shown here:

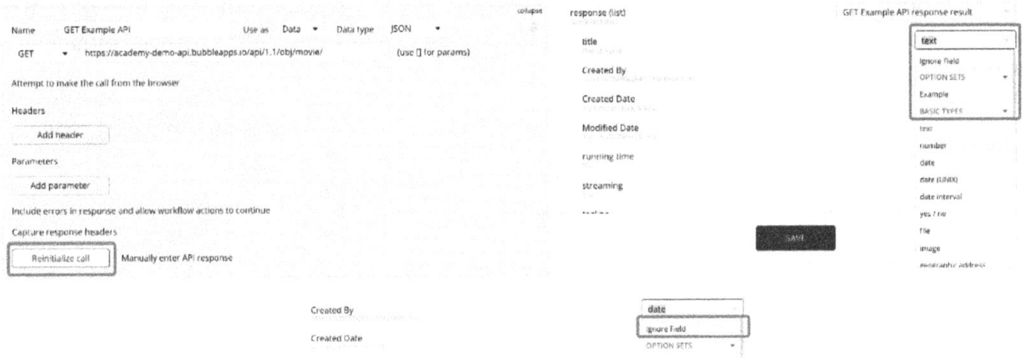

Figure 9.20: API call field types

If there is a field you are not going to use, just click the dropdown and choose the **Ignore Field** option; this will help you get more organized and simplify the API call. That is it – your API is configured, and now you can start using it to build amazing apps with no-code.

Using the API data

There are two ways to use an API after you configure it - Data and Action, let's see how they work in practice.

Data API call

This type of usage will allow you to bring data from the API inside your layouts and UI components by using the **Data source** field inside the properties panel and commonly added by the **Get data from an external API** option. In this example, we are going to use this test API from Bubble, found at the following link: `https://academy-demo-api.bubbleapps.io/`.

The API endpoint we are going to use for this call is as follows: (`https://academy-demo-api.bubbleapps.io/api/1.1/obj/movie/.`)

1. The first step is to create a new empty page, add a repeating group to the page, click on it, and access the properties editor. There, click to configure the type of content as shown in the following figure.

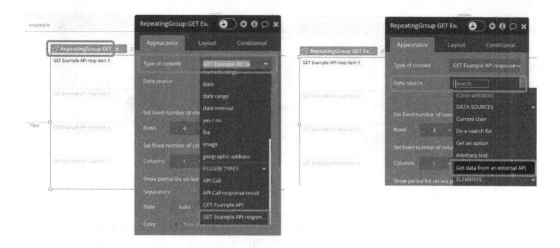

Figure 9.21: Configuring RepeatingGroup

For the **Type of content** option, select your API call name, but select the response option.

2. Go to the **Data source** field, click on it, and choose the **Get data from an external API** option, as shown in the preceding image. Then, continue by selecting the API provider under the dropdown as shown in the preceding image. Choose your API call name.

Figure 9.22: Choosing a data source

3. Next, continue writing your query expression, click **More…**, and choose the **response** option. Now, click outside to clear the selection. You are done configuring the repeating groups' data source using data from the API.

4. Now, you have to define how the elements will show the data coming for the API; create a text field inside the first repeating group cell and click on it to expose the properties panel. Click to insert dynamic data and select the current cell's [name of your API call], as shown here:

Figure 9.23: Dynamic text using API data

5. Continue writing the expression and select the type of information you want to bring from your API call response; in this example, it is the title. Click outside to finish writing the expression and it is highlighted, meaning it is properly configured, as shown here:

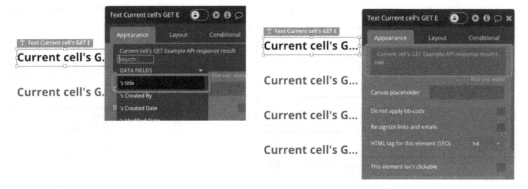

Figure 9.24: Writing a query expression

Now, load your page by clicking the **preview** button at the top and the information from the API should show up on your repeating group. You are done configuring the elements to display data from the API – the same concepts will apply in various scenarios, so play with it and keep exploring!

Another way is to configure your API to be used inside **Workflow** as an action that will be part of your logic while performing certain actions.

> **Note**
> If you wish to use the same API call as data and action, make sure to create it two times, one with each configuration properly set up.

Action API call

This type of API call runs on the workflows, as the name suggests, it is an action. Let's see how to locate the APIs you have configured under the API Connector plugin inside your **workflow** tab. First, head to the workflow tab and click to add a new workflow event; it can be any event or, specifically, it can be an event coming from an action on a page component, for instance, a button.

Selecting the API call in **Plugins** inside **Workflow** as shown here:

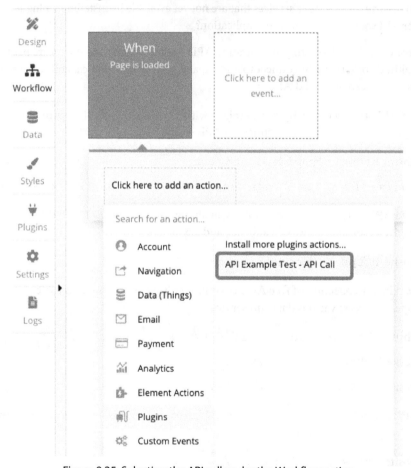

Figure 9.25: Selecting the API call under the Workflow action

> **Note**
>
> If you happen to change an API call configuration from **Data** to **Action**, remember to click the button to reinitialize the call. It is a good practice to use this button if you change settings on the API Connector plugin to make sure everything is working. If you don't set up the API call as **Action**, it won't show on the **Workflow** tab.

Now, once you have the event created, click to add a new action step. You can search for the name of your API call or go to the **Plugins** section and find the name of your API call there, click it, and configure its settings. That is it!

Exploring popular APIs and finding suitable ones

In the previous sections, you learned about plugins and how to use APIs in multiple ways, but what if you want to discover or use other services that are not yet available by installing a plugin? How can you find other API sources to use on your application?

There are websites dedicated to listing and curating APIs – some are free and others are paid, but you can find a multitude of services to connect to and expand your application capabilities: here is a list of a few websites you can go to find APIs:

- **RapidAPI Hub**: An extensive marketplace with a wide array of APIs, providing tools for developers to explore, test, and connect to APIs easily

    - **Website**: `https://rapidapi.com/hub`

    - **List of free APIs**: `https://rapidapi.com/collection/list-of-free-apis`

- **NoCode API**: This offers a range of APIs, enabling users to connect with different services, such as databases, social media platforms, and payment gateways

    - **Website**: `https://nocodeapi.com/`

- **Public APIs**: A collection of free APIs covering multiple industries and use cases, allowing developers to access various data and services

    - Website 1: `https://publicapis.dev/`

    - Website 2: `https://publicapis.io/`

    - Website 3: `https://publicapi.dev/`

- **AnyAPI**: This offers a consolidated view of APIs from different sources, providing developers with an easy search interface

    - **Website**: `https://any-api.com/`

- **API List**: A curated list of APIs in various categories, allowing users to discover and access APIs based on their needs

    - **Website**: `https://apilist.fun/`

- **Postman API Network**: This provides a directory of APIs for building applications, enabling users to search, test, and access APIs directly within the Postman tool

  - **Website**: `https://www.postman.com/explore`

- **APIs.guru**: An open-source database and repository for APIs that provides detailed information, specifications, and links to different APIs

  - **Website**: `https://apis.guru/`

- **APILayer**: An API marketplace that provides a collection of APIs that cater to different functionalities, including data validation, automation, geolocation, and PDF generation

  - **Website**: `https://apilayer.com/`

The following is a list of tools worth mentioning while using APIs:

- **Postman**: This tool helps developers design and test APIs easily. It allows users to create, manage, and test API requests in a straightforward interface, making API development more efficient.

  - **Website**: `https://www.postman.com/`

- **Swagger**: This is known as **OpenAPI Specification**, which offers tools for creating and documenting RESTful APIs. Swagger UI, a part of this toolset, generates easy-to-understand documentation for APIs, enabling developers to visualize and interact with APIs more effectively.

  - **Website**: `https://swagger.io/tools/swagger-ui/`

Using APIs will expand your possibilities while using no-code. With this list of resources, you will be well equipped to find the best APIs to use on your next projects and build amazing applications with no-code.

Summary

In this chapter, we've covered plugins and APIs. You learned about the diverse Bubble.io plugin ecosystem, exploring a curated list of available plugins that can amplify your app-building capabilities. Additionally, you gained practical help on how to install, configure, and use plugins.

This chapter also explained how to choose the most suitable plugins and APIs for your application's unique requirements.

You also learned what APIs are and how to use them to connect with external services and data sources. This chapter helped you gain a strong foundation to be able to use plugins and APIs, empowering your applications to access and leverage external functionalities and resources beyond what is available natively on Bubble.

Understanding these concepts will be very important as you progress further in your Bubble.io journey. In the next chapter, you're going to learn how to deploy and launch your app.

10

Testing and Debugging Strategies

This chapter is a comprehensive guide to help you with the essential techniques to test and debug your application. Throughout this chapter, we will explore the tools available inside Bubble, alongside debugging methodologies and testing strategies.

You will learn how to identify and fix issues before your app deployment to guarantee you are shipping an error-free product into the world.

Understanding these debugging techniques will significantly enhance your capability to pinpoint problem areas and facilitate a smoother development process.

Next, we'll navigate testing strategies so that you can thoroughly evaluate and troubleshoot your application's functionality and performance. Proper testing reduces potential bugs, improves user experience, and builds a more reliable application overall.

This essential step not only enhances your app's performance but also contributes to a professional and credible user experience.

In this chapter, we're going to cover the following main topics:

- Presenting the issue checker and debugger
- Testing your app's functionality and performance
- Troubleshooting and handling errors
- Extra debugging and testing strategies

Presenting the issue checker and debugger

Bubble provides a set of features to help you identify errors while building your application. Most of the time, when you see something in red while creating a query or adding dynamic data to a component, you will notice that there is an error in the way you configured it and you will be able to promptly fix it. However, sometimes, this might not be so obvious and you will be collecting errors while building your app and will only notice them at the end.

One of the most used tools available inside Bubble to help you identify errors is the issue checker. This feature is located at the top bar and it will always indicate if there is any error going on with your build.

The issue checker inside Bubble is shown in the following screenshot:

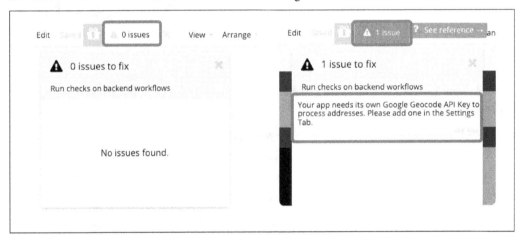

Figure 10.1: Issue checker

The issue checker will help you in identifying and fixing issues. Once it identifies a potential error or missing configuration, a list of issues becomes visible. The number of issues will be always visible at the top bar. When you click the issue checker area, it will open a new window showing a list with the description of each issue you need to fix. Clicking on a specific issue will navigate you to the problematic element or action for correction, helping you find where you need to fix it.

> **Quick tip**
>
> When utilizing the issue checker, consider the following best practices.
>
> First, resolve issues promptly, as soon as you see them. This will help your development process because you might forget things after a while and fixing a lot of errors at the end can be a painful process. Remember that your deployment is restricted until all problems are addressed, so you won't be able to publish your app unless you fix all the issues.
>
> Also, note that Bubble conducts thorough behind-the-scenes computations to identify issues. In larger apps, additional issues on a specific page may become apparent only upon accessing it. It's advisable to navigate different pages to ensure the issue tracker captures all anomalies before implementing app-wide changes.

The second very helpful troubleshooting tool inside Bubble is the debugger. This tool is only visible when you preview your application and it runs on top of your web page as you navigate it. It allows you to navigate your app while identifying potential problems and getting more information about each component and workflow.

To use it, when you hit **Preview** at the top bar to check out your app in action, the debugger kicks in automatically. It stays fixed at the bottom of your page.

Preview mode with the debugger active is shown here:

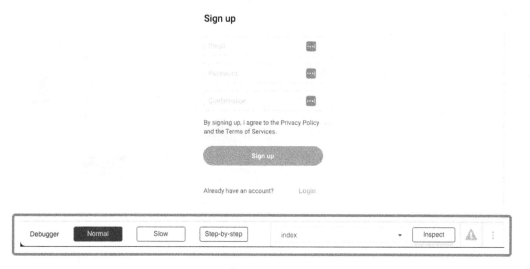

Figure 10.2: Debugger

While you navigate your page, the debugger will follow and allow you to use it to perform an error inspection. When the debugger is on, notice that the URL on your browser will display `debug_mode=true` at the end. So, your full URL might look something like this:

`https://my-bubble-application.bubbleapps.io/version-test?debug_mode=true`

If you want to preview your app without the debugger, to turn it off, all you need to do is tweak the URL parameter called `debug_mode`, remove it from the URL, and hit *Enter* to load the page again. Keep in mind that the debugger is exclusively available for the owners of the application, meaning this will only work for your own projects that live inside your Bubble account. If you try to add this URL parameter on a project that is not owned by you, but still built with Bubble, it won't work, as you don't have access to the editor. Also, notice that the debugger is desktop-friendly and not built to test your app on mobile. To use the debugger, you can click the **Inspect** button. Once you hover over your page element, it will display a red area. If you click that area, you can visualize more information about that element and check for more details or errors.

The debugger with an element selected via the inspector is shown here:

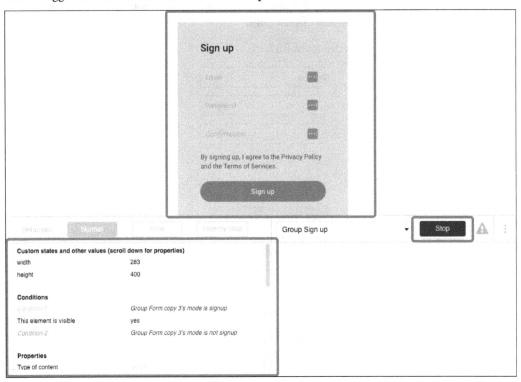

Figure 10.3: Debugger inspector

After inspecting the elements, if you want to exit this mode, just click the **Stop** button; your page navigation will go back to normal. You can also use the drop-down element next to the **Inspect** button to navigate to other elements while inside inspection mode. When performing an action that has workflows you can use the **Normal**, **Slow**, or **Step-by-step** buttons to spot errors related to actions. For instance, if you select **Step-by-step** and click a button that triggers a workflow, you will see each step of your workflow running one by one. This allows you to identify if there is one that is broken or contains errors. If you are a web developer, this tool will resemble the browser inspection tools. We will talk about these later in this chapter.

The issue checker and debugger in Bubble.io serve as indispensable tools for your no-code development while using Bubble. These tools provide a comprehensive overview of potential issues, allowing you to identify and resolve errors efficiently. Play with these tools and get familiar with them; over time, these can become your best allies while troubleshooting errors. In the next few sections, we are going to dive deeper into testing and debugging techniques.

Testing your app's functionality and performance

Building is fun; testing, not so much. However, it is as important as building. After all, you are building something to solve people's problems, not to bring them more.

Testing involves running through various areas of your application to ensure it functions as intended. The goal is to spot problems and fix them before they get into the hands of your customers. Let's explore a few strategies and tools for testing your app's functionality and performance.

Manual testing

The simpler way to run a quick test on your app is by doing it yourself – that is, opening the application in preview mode and acting as a user. You don't need much to do this – you can do it yourself, and this already allows you to spot errors that need to be fixed. You can also create a new account and act as if you were a person – a specific type of user with a goal in mind trying to use the app. Does it work?

Another possible way to test your app is to run your application as a different user, in case you already have any, using their accounts. Once you are inside the **Data** tab, click the internal tab called **App Data**. On the sidebar listing all your database collections, find the **All Users** database collection. Once you've selected it, you will see a table in the middle of the page. This table shows all existing records available inside that particular selected database – in this case, there's just **Users**. If there are registered users in your application, you will see a list of records populated on that database table. On each table row, there is a particular cell with an option called **Run as**. Once you click it, it will open your application in preview mode, running as if you were logged in with that specific user. This feature is useful when you need to test your application from the perspective of other users.

The **Data** tab's **Run as** feature is shown here:

		Email	Address	Admin
	Run as →	email5@email.com		
	Run as →	email4@email.com		
	Run as →	email3@email.com		
	Run as →	email2@email.com		
	Run as →	email@email.com		

Figure 10.4: The Data tab's Run as feature holder

A good idea is also to have multiple accounts for different users with different data so that you can visualize multiple scenarios that could occur in a real situation. Remember, testing your app is the action of viewing your app to uncover possible errors; any strategy that can help you get there is valid.

A second layer of manual testing is creating some sort of script (steps) or documentation to follow during a testing phase. Map and write down the most important use cases and user flows of your app so that you can create a checklist. While testing it, go through it as if you were analyzing your app, but in an organized way.

Here are a few extra tips for you to consider when testing your application:

- **Plan and organize your tests**: Break down your testing into manageable and small pieces. Systematically go through all pages and features to identify issues from time to time. Organize your testing approach according to your development workflow. This way, you can find what works best for you.

- **Create documentation**: Keep focused on the current task during testing. If you encounter an issue elsewhere, make a note for later and add it to your documentation. Staying organized helps you avoid missing potential problems. It also helps when working with more people since you can share information.

- **Test with real data**: Populate your app with test data. An app's behavior can vary with different data volumes, revealing potential issues related to design, performance, and security. Real data is the true test if it will work because users are unpredictable.

- **Test your app on multiple devices**: Ensure your app's compatibility by testing it on various screen resolutions and devices. Use tools such as Chrome Developer Mode to simulate different screen sizes and connection speeds. Use real devices if possible to feel the experience of your application in different scenarios.

- **Role-based testing**: Test your app from different user perspectives. Introduce privacy rules and conditions to simulate how users with different access levels experience the app. For instance, test as both a regular user and an admin to catch potential issues or inconsistencies.

Turning testing into a process can take your application to another level, making it more organized and preventing you from forgetting to test specific areas of your app. This can help you focus on the most important features that have to be tested and help you release applications with fewer possible errors to your users. It will help you remember to test the key features that must work correctly and deliver a good experience to your users; this is essential for your app's success. Now, let's see other useful ways you can test your no-code application.

User testing

Another way to test your app is by putting it in front of real users – it could be friends or at least other people who are not building the product. Since you will spend a lot of time building it, you can create some blind spots or bias, and maybe won't notice a few things over time, so getting someone else to help you can quickly become a valuable way to test your application. If you can find and talk to your actual target users and ask them to test it in front of you, then you have an even better scenario, so do that if you can. You can explain that the application is still under development and it can potentially break, but that is your goal – to test it and find errors. An alpha or beta release can also become a viable option during the test phase.

When testing with people, it is important to do so in a more professional and organized way. You can do it roughly as that's better than not doing it at all, but if possible, apply some of these tips:

- Prepare before testing; plan what you are going to test and the specific situations your user will go through. If you are going to record it, make sure the setup is ready and working. Test your setup and tools.

- Set meetings with space between them so that you have time to prepare for the next session and take a little break. If your test is going to happen remotely, make sure your internet connection is stable and that the person on the other side has everything needed to participate in the test.

- Invite people for the test, online or in person, and make sure you confirm their participation a day before or a few minutes before the event to avoid no-shows.

- Once you start the test, let the person know you are not testing them but your application. Then, explain the situation and be upfront by saying that you are testing your app and that you are happy they are going to help you test it. Thank them for their time and participation. Say that there is no right or wrong; you just want them to behave naturally and don't worry if they don't know or don't understand something. Make sure the users are comfortable and not worried about being tested. If you don't have experience with user tests, you can create a document and write a script and steps to help you. There's no problem reading during the test – you want to make it right, so use a document if needed.

- During the test, ask the person to say what they are thinking. By doing so, you can try to understand their thoughts, questions, and behavior.

- You can choose to record the session. If you do, ask the participant if that is ok; if they don't want to, just accept it and move on. If they do accept, just make sure that this won't interfere with the tests.

- Don't take too long! Tests should be simple for your users, so don't add too many complex steps, and don't ask testers to spend more than 1 hour testing your application. If you need to test multiple things, break them into smaller steps and do specific tests for each part. Usually, you are going to simulate a particular scenario and create specific tasks. During the test, ask the users to perform these tasks and see if they can accomplish them or not. That's the idea.

- Take notes and, if possible, ask someone to help you. Testing with two people is more efficient – one person can lead the test and guide the tester while another can be the note taker and be free to type anything that is needed without worrying about another task. Another tip is to prepare a file or form before the test so that the person taking notes can do so in a more structured way.

- If people are not volunteering to test your app, you can give them something in return, such as an Amazon gift card or something else that might be of interest. This is a common way to attract testers and also say thanks for their time.

With these tips, you are more likely to run great user tests. The more you do this, the better it gets. Keep practicing!

Another way of testing your application with real people is by submitting it to testing websites and paying people to test it for you. Websites such as `https://www.usertesting.com` can be an alternative.

Remember that Bubble has a huge community and you can also make friends and ask people to help you and test your application.

Performance testing

Performing manual tests and user tests can help you spot usability issues, layout problems, workflow errors, and much more. It can help you notice if your application is slow and has performance issues as well, but if you want to evaluate and test the performance of your software, you can use specific tools that were built for that. This can give you way more insight and information about what needs to be improved.

Here is a list of tools you can use to run performance tests on your application:

- **Google's PageSpeed Insights**: Analyzes the performance of web pages and provides suggestions to optimize loading times: `https://pagespeed.web.dev/`

- **Gmetrix**: Offers insights into a website's performance and provides recommendations for improvements: `https://gtmetrix.com/`

- **Semrush**: A comprehensive SEO tool that provides insights into website traffic, keyword analysis, and competitive research: `https://www.semrush.com/`

- **Ahrefs**: Focuses on backlink analysis, keyword research, and competitor analysis for SEO purposes: `https://ahrefs.com/`

- **SEO SiteCheckup**: Offers SEO analysis, monitoring, and tools to improve website performance: `https://seositecheckup.com/`

- **Google Lighthouse**: An open source tool for improving the quality of web pages by auditing web apps for performance, accessibility, SEO, and more: `https://developers.google.com/web/tools/lighthouse`

- **Wave**: Evaluates web content for accessibility issues by analyzing the page and providing human-readable results: `https://wave.webaim.org/`

- **UsableNET**: Offers automated and manual accessibility testing tools to ensure web content is accessible to all users: `https://www.usablenet.com/`

- **Browserstack**: Provides a cloud-based platform for cross-browser testing, including accessibility testing: `https://www.browserstack.com/`

- **Responsinator**: Displays how a website looks on various devices, helping assess its responsiveness: `http://www.responsinator.com/`

- **Screenfly**: Allows users to view a website on different devices and resolutions, aiding in responsive design testing: `https://screenfly.org/`

- **Chrome Responsive Viewer Extension**: An extension for Google Chrome that provides a quick way to view a web page on various device resolutions: `https://chromewebstore.google.com/detail/responsive-viewer/inmopeiepgfljkpkidclfgbgbmfcennb`

With these tools, you can test a wide variety of aspects of your application. It is up to you how extensively you want to perform tests, but the more professional you become, the more important it will be. With these tools, you can test page speed and overall performance, improve SEO to rank higher on search engines and help the marketing team, test your app in different browsers, check accessibility issues, check color contrast, and preview your application in different sizes and devices to evaluate if it is working responsively. Now, go test it!

Professional testing

If you have a bigger budget, you can hire a no-code developer to test it for you, run an audit, and even fix the errors for you. You can also hire QA companies that focus on testing software and providing a full report on what to fix. Companies such as Flusk do a full test of your application and can also provide you with a security report and help you fix these issues. There are also agencies and freelancers out there that can provide testing and support or even application analysis for you on demand.

Now that you've learned about various methods and tools you can use to test your application, let's learn a little bit about how to troubleshoot and handle errors once you find them.

Troubleshooting and handling errors

Building software is going to require you to handle bugs and errors. You will typically find them after the testing phase. At this point, you have to fix them; this isn't great, but it's part of the job. Some never do and call it a technical debt. No matter what you decide to do, errors will exist, and fixing them is important. In this section, we are going to talk about possible ways of troubleshooting and handling errors within your Bubble.io application.

Troubleshooting is an art; it requires patience and a process, and you will develop your own over time. When you find yourself in a situation where there is an error or something is not working, it can cause you to feel different types of emotions. After facing several errors and not being able to fix them, this can lead to anger and frustration. These feelings make everything even more difficult.

If you find yourself on a river of bugs, one very simple but effective tip is to *go back to the basics*. What does this mean? Well, sometimes, you have to unbuild things to fix them. This means it is better to go back a few steps and redo it in another way than trying to fix it.

Trust me – sometimes, the problems are at a stage where everything is too complicated. In this case, start to untangle things by removing what isn't necessary and starting things from scratch. Another strategy is to try to build or replicate just one specific scenario on a new and blank application. This way, you remove the chaos and just focus on what you are fixing, and you also remove the chances of having conflicts or other things impacting what you are doing. So, remember, if things are too complicated, go back to the basics.

Another tip is to run tests frequently and don't wait too long to discover issues – the quicker you spot them, the faster it will be to solve them. Avoid building too much and too many new features without testing them at an initial phase. Remember to always simplify your project scope; when projects get too big, testing becomes more difficult.

Removing plugins can also be a good idea. This way, you can test your application without external influence or potential conflicts. Bubble has a feature called safe mode. It provides a preview of your app while temporarily disabling specific elements for debugging purposes, including on-page HTML elements and community-made plugins. Enabling safe mode allows you to identify issues introduced by plugins or custom code. To activate safe mode, press and hold the mouse button on the **Preview** button for one second; a dropdown will display the available options.

We can see the various safe mode options here:

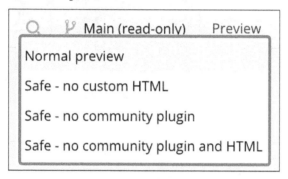

Figure 10.5: Safe mode

Activating safe mode will allow you to remove distractions and potential conflicting code that may be running on your application. If you entered safe mode and the errors are gone, you already know that you should look for plugins or code that may need a fix. This can save you a lot of time and help you fix problems with your app.

Communication can also be a problem while handling errors. Do you understand what the error is and how to replicate it? If users are giving you information about an error and saying they found something that isn't working, talk to them, try to see it for yourself, and understand what's going on before even trying to fix anything. One very important thing is to understand the error and be able to replicate it. Something we won't be able to replicate is errors on our end. This can be frustrating because we won't be able to test and see them in our environment. So, make sure you understand the error and can replicate it before trying to fix it.

Server logs provide insights to help you effectively debug your app. While the debugger allows you to test and debug the current state, the server logs offer a historical perspective, allowing you to investigate past issues.

To access the server logs, follow these steps:

1. Click on the **Logs** tab.

2. Navigate to the **Server logs** tab:

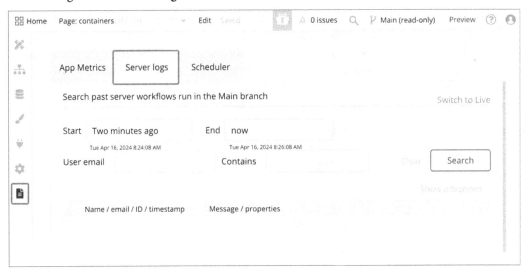

Figure 10.6: Server logs

On the **Server logs** tab, you can review changes that were made during testing and understand the sequence of events when a live user encounters a problem. It's important to note that the server logs for development and live environments are different, so make sure you are working in the right environment. Custom branches, if any, are encapsulated within their respective development and live environments.

The server logs provide a detailed history of backend activity, such as database updates, button clicks, or page loads. You can filter logs by date range, errors, specific users, or text strings. The logs are presented in reverse chronological order, each row representing a separate activity. Server logs can help with troubleshooting and understanding app behavior, making them valuable resources for both developers and those filing bug reports.

It is also possible to store custom logs directly in Bubble's server logs for efficient error tracking. For instance, when creating an API workflow, you can add fields called "error_log" that allow you to log errors without storing them in the database. The workflow takes parameters such as "code" and "message," making it flexible for various error scenarios. A sample page with a button can trigger this workflow to create a sample error log. This method can be used for tracking errors in real time without cluttering the database. After storing these logs, you can search and view them directly in the server logs, enhancing the debugging process.

> **Quick tip**
>
> Leverage help from your users to learn about bugs, errors, and potential improvements. Create a direct communication line with them – it can be a simple contact email or a custom form that allows users to reach out to you. Don't forget to ask them for their contact information and screenshots or videos about the error so that you can easily see and understand what they mean. Sometimes, replicating bugs can be a hard task because you work in a different environment/computer. That's why you need visuals. If you want to go further, implement user feedback tools in your project, add a feedback button, and allow your users to provide more insightful feedback more easily.

If you have a lot of errors to fix, you can create a list and prioritize them by order of importance. This way, you remove pressure knowing that they will be fixed but not at once. Adding an organization to your process can help a lot and prevent you from making your life a nightmare. Having a sane head will also help you fix errors with more ease. When working on a problem for too long, if you are still stuck, take a break, stop for a while, talk to someone, and give yourself some time to digest the problem. Often, when we go back to the development process after taking a break, things change and we may find the solution in just a few quick, easy steps. So, allow yourself to breathe; this can help sometimes – give it a try. Be patient! Problems are problems – we can't change them or stop them from happening, but with patience, we can find a way out. With this mindset in place, let's consider the next set of tips and strategies you can implement to continue improving your application. This time, we'll look at some extra tips and tricks for testing and debugging your application.

Extra debugging and testing strategies

In this section, we are going to explore additional techniques and tools to help you find errors and debug your application. As we saw previously, Bubble provides tools and features to help you test and find potential problems while developing your app, but there are other ways to proceed with advanced debugging techniques to help you uncover insights that empower you to take your development process to the next level. We'll first with some options that are still available inside Bubble. After I will show you some external tools you can use.

Display options and responsive mode

Inside your Bubble editor, you can turn some features on and off so that you can better visualize and understand what's going on with your layouts. On the top bar, there is an option called **View**. When you click it, you will see a few options that allow you to customize how your editor canvas will look:

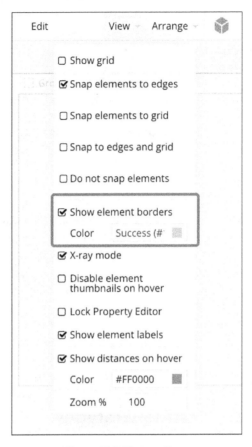

Figure 10.7: View options

For testing and debugging, two options can be useful:

- The first one is **Show element borders**. Here, you can pick a color and all your components will have a border with the chosen color. This can help you visualize components and see how they interact with each other.

- The second useful option is **X-ray mode**. Once you turn it on, it will reveal the components in layers and add some transparency to all the components that have been added to your page. This allows you to see if there are any components on top of each other or if components are overlapping.

Another useful area to test and adjust your layouts is responsive mode. There, you can quickly visualize and test different viewport sizes. Combined with the previous techniques, such as **Show element borders** and **X-ray mode**, it can become a powerful method to find potential problems and fix them right inside Bubble.

Here's an example of the responsive editor exposing element borders:

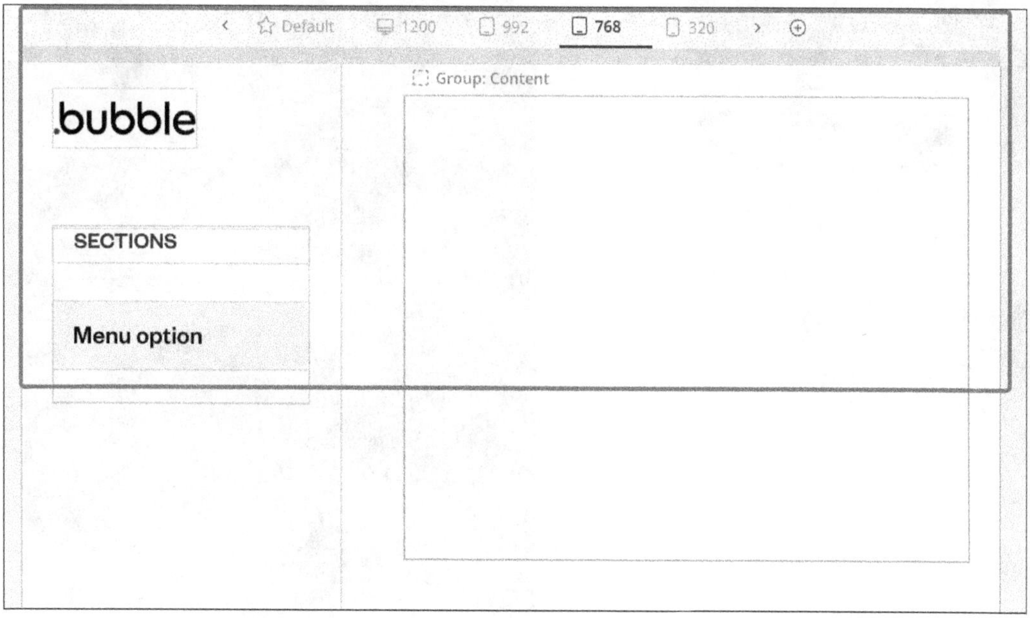

Figure 10.8: Responsive mode

These are simple techniques that can help you spot potential problems. Since you are still inside Bubble, you can quickly navigate between pages and fix your layouts without leaving the tool. Now, let's continue and look at other methods that can be used to test your application, but this time outside Bubble.

Testing with browser tools (Chrome Dev Tools)

Your browser is one of the best places to test and debug your app because it is the actual environment where your app will be displayed. Chrome has a native browser inspector that is very powerful and can be a helpful tool to test your application. It's called Chrome Dev Tools. If you already use Chrome as your browser, you already have it, so don't need to install anything extra. To use this tool, you just need to go to the website you want to test and hit *Command + Shift + C* (macOS) or *Ctrl + Shift + C* (Windows) on your keyboard.

The following screenshot shows what Chrome Dev Tools looks like:

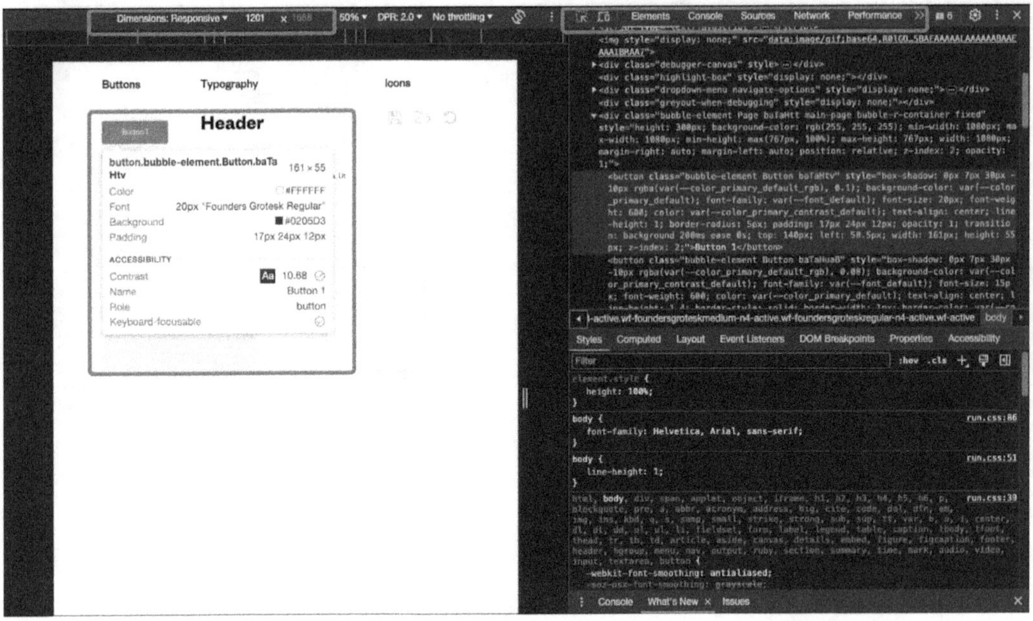

Figure 10.9: Chrome Dev Tools

Other browsers have a similar feature, such as Firefox, but I am only going to talk about Chrome here. If you use other browsers, consider using Chrome, at least for testing and using Chrome Dev Tools. Once you hit the keys to open the inspector, you will see your browser window change; it will expose the code of the website on the side and many other tools you can use to start testing and debugging your app.

Now, there are a lot of tools and features available and some are more advanced. If you are really into it, I recommend that you conduct some research or even do a course about it. I'm only going to cover the basics here.

On the left-hand side, you will see your page. At the top of the window, there is an area where you can choose to make the preview responsive; you can drag on the sides of the page to stretch it to any size or you can pick which types of device you want to simulate.

On the right-hand side, you will see the code. Here, you can find the HTML and CSS of that page and even change it to preview how it would look. You can use the console to check JavaScript and the **Network** tab to see your page files loading. For more advanced users, you can use it to spot security issues and much more.

If you hover your mouse over the top of a component, you will see information about it; you can select it and even delete it from the page if needed. Imagine this as a preview/editor but outside Bubble. It works for any web page and is very useful. Give it a go and start getting familiar with it.

No-code-specific testing and monitoring tools

The no-code space is maturing. More and more tools are available nowadays, and we as no-code developers are also getting more professional and becoming more specialized, we are incorporating techniques and processes that come from the development world. This is a good thing and I believe it will continue to evolve.

A few years ago, we barely knew what no-code tools were, and there were just a few tools available. Today, we not only have plenty of tools to choose from but also tools and resources dedicated to helping no-code developers who use and focus on a specific no-code tool – in this case, Bubble. The Bubble environment has grown significantly, with a global community of developers and companies around it. Agencies, freelancers, companies, and startups building with Bubble is now a reality, and as this environment and the Bubble ecosystem grow, it will become even stronger. At the same time, professional developers have already jumped into no-code and low-code, and even enterprises and big companies understand the value of using such tools to speed up their development while reducing costs.

To help you with this journey of navigating the Bubble ecosystem and professional tools and resources available to help with making no-code development even more sustainable and professional, we'll look at two tools that can help you test and monitor your applications. I hope they can help you during this process.

NcScale

This is a monitoring service that's tailored for your no-code stack, offering a suite of features designed to simplify troubleshooting, enhance visibility, and instill trust in your infrastructure. The key features of ncScale include dependency tracking, versioning, full-text search, obfuscation, log management, alerting, an assets catalog, and an informative overview. This service ensures efficient monitoring of your entire stack, providing valuable insights into dependencies, facilitating version control, enabling comprehensive search capabilities, and safeguarding sensitive information through obfuscation. Additionally, ncScale offers tools integration, allowing seamless connections with platforms such as Xano, Zapier, Make, Weweb, Baserow, N8n, Airtable, Bubble, custom code, and more in just one click. With its user-friendly interface and powerful features, ncScale empowers users to monitor, troubleshoot, and optimize their no-code applications effortlessly.

Website: `https://www.ncscale.com/`.

Flusk

Flusk is a cutting-edge security and monitoring tool specifically designed for Bubble.io that offers advanced features to enhance the safety and performance of your applications. Trusted by top-tier companies, Flusk stands out as a comprehensive solution chosen by leaders in high-growth startups and enterprises worth over $400 million. On the security front, Flusk automates over 20 Bubble-specific security checkpoints, covering areas such as Data Leaks, API Workflow Protection, and Page Access. It conducts security audits upon deploying a new app version, providing instant results to keep your app secure effortlessly. Flusk's documentation ensures you understand and address identified security issues. In the realm of monitoring, Flusk allows you to track and remedy errors that users may encounter, providing a simple dashboard to visualize and address issues. The tool automates security checks for new deployments and offers smart deployments, allowing you to roll out versions when users are not online. Flusk's Visual Logs provide a modern and user-friendly approach to exploring and debugging your Bubble logs. With Flusk, users can activate more users, boost feature adoption, and drive expansion revenue, making it a robust solution for securing and optimizing Bubble.io applications.

Website: `https://www.flusk.eu/`.

These tools are just a few of the resources available out there. You're encouraged to keep researching and finding new solutions created by Bubble developers for Bubble developers. I am sure that at this exact moment, a few folks are working hard to build even more powerful solutions to help you on your Bubble and no-code development journey. The Bubble community is an ever-growing community, just like the no-code movement and space are. You can leverage the power of community, as we will see in the next section.

Community and support

I want to emphasize the power of community. Remember, Bubble has more than 3 million users (about the population of Arkansas) out there and is growing every day. When you get stuck or need help, rely on the community to help you test and debug your app. Go to the forum and post on groups and social media. Someone will help. If needed, hire a professional, ask for support, mentorship, and consulting, and get unstuck.

You don't have to be alone. Remember, there is a community of no-code developers like you out there.

Testing and debugging can be painful for some and fun for others. The more experienced you get, the less it will be a problem. But problems are still problems. After all, we are all learning, and no-code is still growing and evolving. It is ok to not know everything. Try exploring and experimenting as much as you can, and allow yourself to fail. We are humans, and we make mistakes. Don't let that stop your passion from building. Keep going! You will find a way out soon and I hope these techniques, tips, and tools can help you along the way. Welcome to the Bubble community. Let's keep building...

Summary

In this chapter, you learned about testing and debugging, an essential aspect of any app development. You learned tips and tools to effectively test your app's functionality and performance, ensuring a robust user experience.

We also presented internal and external tools to help you troubleshoot and handle errors that can unexpectedly arise during development.

Additionally, we explored extra debugging and testing strategies, providing you with a comprehensive toolkit for creating resilient and reliable applications. Understanding these concepts is important for building high-quality apps, and the skills acquired will play a crucial role in your continued learning journey. In the next chapter, you will learn how to deploy and launch your app.

11

Deploying and Launching Your App (Publishing)

Congratulations! You've spent countless hours crafting your application, and now it's time to present it to the world. This chapter will guide you on how to deploy and launch your project to ensure a smooth and successful deployment.

By the end of this chapter, you'll possess the essential skills and knowledge necessary to prepare your app for deployment effectively and confidently navigate the process of launching your app to the public. You'll understand the significance of meticulous preparation, ensuring a smooth and error-free launch, and be equipped to make your app globally accessible.

We will cover the following topics:

- Understanding the concept of deploying an application
- Preparing for deployment – reviewing, previewing, and testing the app
- How to set up a custom domain
- Launching your app to the public

Understanding the concept of deploying an application

After building your application, it is time to make it live and available to the world. This is a very exciting moment, but it also requires paying attention to avoid problems. If you are new to the tech world, you might be wondering, what does it mean to deploy? Deploying an application refers to the process of making the developed software or application available and accessible to end users. Usually, when you develop software, you run and test it on a separate environment, which is only visible and accessible to the development team. This project is not released to the public until it is ready. Once it is ready, you send it to the outside world so that everyone can see it. Bubble has two different environments: the development environment and the live environment. These are separate.

When you start building things inside Bubble, you will be working on the development environment and will have to manually choose when to send things to the live version.

An example of your Bubble app's environments is shown here:

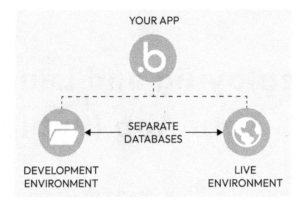

Figure 11.1: Bubble development and live environments

Simply put, deploying a Bubble application means transferring or updating the current version of your application from the development environment to the live environment, making it accessible to end users.

> **Note**
>
> You will only be able to deploy an application when signing up for a paid plan. The free plan only allows you to build and preview the application using a free subdomain that displays the application version running on the development environment. To add your custom domain and deploy your application to the live environment, you will have to upgrade.

It is good to know that your application also has two databases, one for each environment (development and live). You can go to the **Data** tab and, under **App Data**, you will see a red link stating **Switch to live database**. Once you click it, you will see the same database but with data from the live environment. If you click it again, it will bring you back to the development version.

Think of the live environment as a copy of your development application that is available to your users, like versions. The development environment is always the original copy that you, as a developer, have access to. Once you are ready to send it to the world, you can send it live by deploying the development environment app version; when you do this, it will be sent to the live environment. Every time you are working on the application and make a few changes to the development environment, it will not affect the live version, it will just change the development version. Once again, if you are ready to send it live, you can do so with a deployment. Remember that the live version will have users using it and generating data, so, before sending data to the live environment from the development database, remember to bring data from the live environment to the development one since the live version will always be the most up-to-date. If you forget to do that, users might lose their data. Bubble also has a version control system that allows you to manage these copies and even revert to one specific version. The features that are available in the version control depend on the plan you choose when upgrading to deploy. If you have a technical background, know that Bubble has a branch area and allows you to create save points for your application.

Once you are ready to deploy, click the **Upgrade to deploy** button at the top of your editor, pick a plan, and follow the checkout process.

Before you deploy an application, it is important to do a project review and prepare for it. You'll learn how to do this in the next section.

Preparing for deployment – reviewing, previewing, and testing the app

Providing a great user experience to your users is key. Deploying a broken app can damage your app's reputation and can be a headache if you are working for a client. So, to avoid these situations, doing a meticulous review before deploying is important. If you are working in a lean and agile way, you might want to deploy and ship new features often, but that doesn't necessarily mean you will neglect testing. It is important to add a final step to your development sprints before the deployment to ensure you are shipping things that work and were tested. If needed, work with a QA team to help you with this process. If you are working by yourself, just make sure to add this to your routine. Before sending your app out into the world, do a proper review and test to ensure it is ready to use and working properly.

Here is a list of some important things to consider when deploying a Bubble application:

- Check and resolve any issues flagged by the issue tracker.
- Check your page layouts and preview and navigate to the different pages to check if they are rendering properly.
- Check if your pages are working on different devices and if they are responsive. If you plan to allow users to use it on a mobile device, check this as well.
- Review all the updates and changes made in this new version.
- Run the main tasks a user would do on your app to check if the core features are still working properly.
- Test and review API integrations and plugins.
- Check your database structure and data. Remember that your live database will always be more updated than the development one.
- Confirm security rules and user access.

> **Tip**
> Deploy your app when the usage of the app is low, such as at night, to minimize the impact on users visiting your site at the time of your deployment. You can also choose to lock your application pages, showing a message that it is under maintenance, thus not letting users create new data and use the app. Just make sure you inform people beforehand via email and let them know about the scheduled maintenance.

If you are working in a team, it could be a good idea to have some sort of documentation to keep track of deployments and allow communication. This can be a good way to avoid problems and keep everyone on the same page.

A good practice is to deploy updates in small batches as this helps with avoiding errors and simplifies things. You will only need to test fewer things and it allows you to ship new updates faster to your users. If you are not familiar with the tech space, I recommend that you check out the Scrum methodology and lean startup concepts. Getting familiar with these will help you develop better software in an agile way.

If things go wrong, don't panic – there is always a way to revert changes. Just make sure you plan things properly and have a backup plan. Fortunately, since you are working with software, it is very easy to change things and fix issues if needed; just make sure you don't take too much time to spot potential issues after the deployment. Having a way for users to communicate with you in case things go wrong can also be a good idea so that they can also help and tell you right away if things break.

The more you do this, the better this process will become, and the more experienced you will be in creating a process with rules, guidelines, and best practices. In the next section, you are going to learn how to set up a custom domain so that your users can access your application using yourcompanyname. com. This can be a more professional way to spread the word about your app rather than using free subdomains provided by Bubble.

How to set up a custom domain

Adding a custom domain provides a professional and branded touch to your application's web address. Let's go through the step-by-step procedure to configure and link a custom domain to your app:

1. The first step is to choose a domain name. Go to a domain registrar website of your choice, such as Godaddy, Google Domains, NameCheap, Name.com, or Porkbun, and buy the domain. You don't need to buy hosting, just the domain. Hosting is already provided by Bubble.

2. Now, go inside Bubble and upgrade your plan to at least the first paid plan so that you can set up a custom domain. Free plans won't let you configure it, so it is important to upgrade your plan.

3. Open your project and navigate to the **Domain settings** tab. Under the **Domain/email** section, you will see a place to add your custom domain. Simply type the domain you bought there:

Domain settings as shown here:

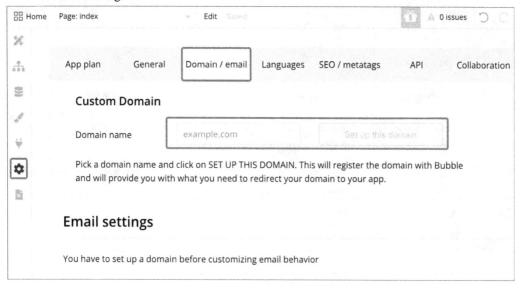

Figure 11.2: Domain settings

4. After typing your domain URL, click the **Set up this domain** button. That's pretty much all we must do inside Bubble. The next part of the setup will happen inside the registrar's website – that is, the place where you bought your domain.

Before you leave Bubble, just pay attention to this new information that's been provided. These are the DNS values we will use to configure the domain.

The DNS information is shown here:

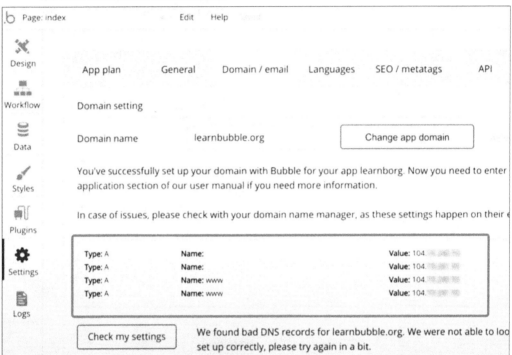

Figure 11.3: Domain DNS information

5. Now, log into your account on the website where you bought your domain and find a page called **DNS**, **DNS Records**, or **DNS Settings**. If you are not sure, you can also refer to their documentation or ask for help from support – say you are trying to configure DNS records and provide the values available inside Bubble.

6. Once you find the page, you just need to change a few values. Delete or update the existing records if there are any available. Create new ones with the values from Bubble. Note that every domain registrar's website is different. In this case, we are going to show an example from GoDaddy's website.

7. The **Records** page is shown here:

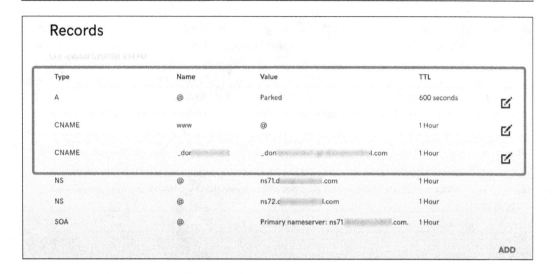

Figure 11.4: GoDaddy's Records page

If you are using GoDaddy, the A and CNAME records will be available by default. Remove these – we are going to add new ones.

8. After deleting the existing records, click to add a new one.

9. The following screenshot shows how to create a new DNS record:

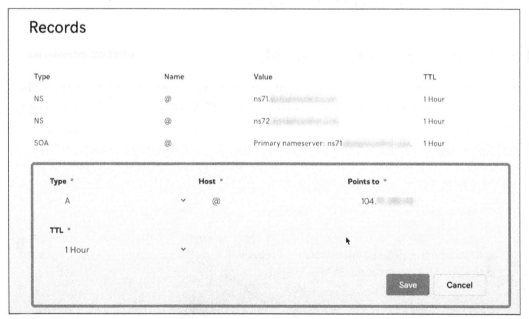

Figure 11.5: Adding a new DNS record

10. Your goal is to create four new records, following the same values available on your Bubble app. The first two records are empty (without the www), while the other two have the www. So, first, create a new A record. For **Type**, choose **A**, and for **Host**, type @. The IP address comes from the Bubble website – that is, the first one (without the www). Do the same with the second. Now, for the ones with www, choose **A** for **Type** again but for the **Host** field, type www, then copy and paste the IP address. Do that one more time with the fourth line and you will be good to go.

> **Note**
>
> The domain setup process can take from 24-48 hours to propagate. This is normal. It can happen earlier but just in case, consider this period before you try something new or contact support for help.

> **Tip**
>
> When buying a domain, choose an option that comes with SSL encryption. This provides your website with more security and increases trust from your users.

Once your domain has been set up, when you click **Preview**, you will notice your Bubble app opening from your custom domain. This means that you're good to go! Now, tell the world about your app and share your website on your social networks so that people can access it.

Now that you know how to configure your custom domain, it is time to launch your app to the public. We're going to do that next.

Launching the app to the public

Once your app has been refined and tested, it's time to introduce it to the world. Launching can be exciting and you can choose how it is going to look. In practice, it just means pushing a button, but you can choose to celebrate and tell the world about it too. It's up to you.

Here's a quick step-by-step guide on how to deploy your application. To deploy your application, you need to upgrade to a paid plan. Click the button at the top to **Upgrade to deploy** and choose a plan. After that, proceed with the checkout process.

Once you have a paid plan, you can start deploying:

1. When your issue tracker is clear, and you're ready to deploy your app, you can initiate the process, which is virtually instantaneous. Begin by clicking the version control button located in the top-right corner of the Bubble toolbar:

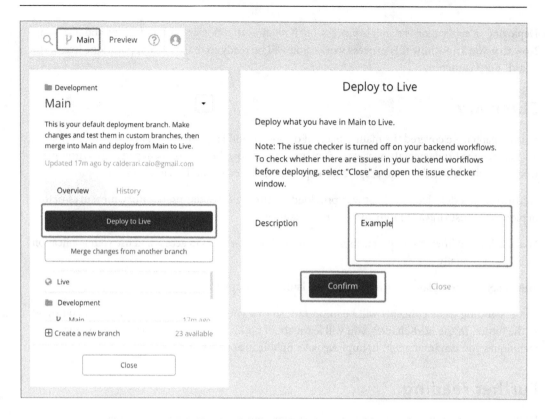

Figure 11.6: Deploying your app

2. After clicking the main button, a new window will open on the side. If any impediments are preventing the deployment, you'll see notifications at this stage. To proceed with the deployment, click the **Deploy to Live** button.

3. After that, a new pop-up window will open. You have the option to describe the deployment to document the modifications you've made. This description is automatically saved with the savepoint, allowing you to revert to an older version in case any issues arise from the deployment. Now, add a small description of this version to help you remember it later. Then, once you are ready, click **Confirm**.

4. And that's it! A new window will show you a success message and give you the URL so that you can test your newly deployed application. This means you can navigate and test it on the live environment. At this point, your users will be able to use and interact with the new version.

It is also recommended that you create communication channels so that you can talk to your users and let them know about the new features and updates that were released, for example. This can help them stay informed about how your application is evolving and becoming better for them.

Deploying an application is a simple process, but it requires attention and prior planning and reviewing. Now that you know how this process works, you will be ready to make your applications live to the world! Keep launching.

Summary

In this chapter, we covered the main aspects of deploying and launching an application on Bubble.io. You learned the fundamental concepts of deployment and how the development and live environments work. You also learned how to prepare for deployment by testing and reviewing your app.

Additionally, a detailed checklist was provided, outlining essential preparatory measures before deployment, ensuring a smooth and error-free launch.

You also learned how to set up a custom domain to allow your users to access and use your application more professionally.

This chapter also explained how to effectively launch your app to the public.

Understanding these concepts will be important so that you can launch any application you build in the future. In the next chapter, you will learn about optimization and performance enhancement techniques you can implement to improve your Bubble.io applications.

Further reading

If you need more information about branches, version control, and deployment features, please refer to the Bubble official documentation: `https://manual.bubble.io/help-guides/getting-started/navigating-the-bubble-editor/deploying-your-app`.

12

Monitoring, Maintenance, and Updates (Apps Governance)

In this chapter, you will learn about app management and how to interpret dashboard numbers to keep track of your app's performance and optimization.

We start by understanding **workload units** (**WUs**) and explore how these units quantify resource utilization and its impact on the app's functionality and operational costs.

Transitioning into the daily life of an app owner, we talk about updates, version control, and app governance, important subjects for any company building no-code applications.

Monitoring your app's performance, maintaining it, and keeping it updated are important steps after deployment. Valuing user feedback and making data-driven decisions according to user experience are key to making sure your app is a success.

In this chapter, we're going to cover the following main topics:

- Understanding dashboard metrics – An explanation of WUs and usage
- App management – Versioning control, updates, and production to Live
- App governance – Defining responsible people in your organization
- Monitoring app performance and user analytics
- Handling maintenance and addressing user feedback

Understanding dashboard metrics – An explanation of WUs and usage

After deployment, your goal will be to analyze what is happening with your application. You will be collecting a lot of information about your users and important data that can be used to optimize your app's performance and user experience. We are going to cover more on that later. In this section, we will focus on the dashboard metrics related to your app's performance and WUs consumption, as this plays a huge part in how much your app will cost to run monthly.

What are workload units (WUs)?

WUs serve as a measurement of the system resources utilized by your app. It is the amount of work that **Bubble** does to power your application. They represent and quantify the computational resources used when your app performs various activity types, such as database queries, workflows, file uploads, API calls, and calculating a total. This metric measures all the underlying activities your app performs as it runs, and the total workload your app consumes is tracked over one month. WUs are a fundamental element in Bubble's pricing structure. The more WUs required to run your app, the higher your app cost will be. Explaining WUs involves understanding how they relate to app functionality and user actions. That is why it is important to understand how WUs work and optimize your app's performance, ensuring efficient resource utilization and managing operational costs.

The **Workload Usage Dashboard** is shown here:

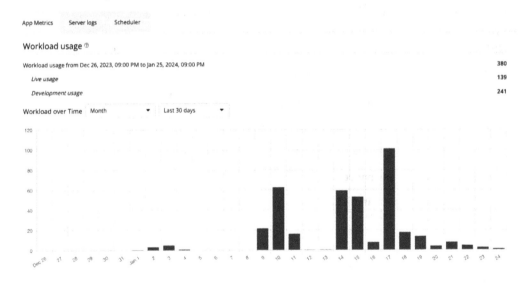

Figure 12.1 – Workload usage graph

Here's a breakdown of how WUs are used:

Workflow execution

Each action within a workflow consumes a certain number of WUs.

The complexity and type of action impact the number of WUs consumed. For example, making an API call or performing a database operation may use different amounts of units.

- **User interactions**: When users play with your app, such as clicking buttons, navigating between pages, and triggering workflows, all these little actions contribute to WUs usage. Dynamic content updates or conditional workflows can also add to WU consumption.

- **API calls**: When your app makes API calls, each call consumes WUs. The complexity and data transfer involved in the API call influence the number of units used.

- **Data operations**: When your app handles data manipulation, such as creating, reading, updating, or deleting records, it will consume WUs. The size and complexity of the data operation impact the WUs consumed.

- **Recursive workflows**: If workflows trigger other workflows (recursive workflows), each iteration contributes to WUs. As a developer, you need to be mindful of avoiding infinite loops that could lead to excessive WU consumption.

- **Scheduled events**: When you schedule actions to run at specific times or intervals, such as recurring workflows, they also use WUs.

- **Third-party integrations**: Integrating third-party services or plugins may involve additional WUs, especially if they require complex computations or extensive data processing.

- **Complex calculations**: Performing complex calculations within workflows or utilizing advanced features may consume more WUs.

Understanding WUs helps developers estimate the computational cost of their app's functionality and make informed decisions about optimization. It's crucial to monitor and manage WU usage to control costs and ensure efficient performance, especially considering the limitations imposed by Bubble's pricing plans.

How can I track my app's workload usage?

You can track your app's workload usage using the **App Metrics** dashboard in Bubble. The dashboard provides several visualizations, including a bar graph that shows the total workload over time, broken down into days and hours. Each column in the bar graph represents one day, and you can click on a day to zoom in and view the hourly consumption. Additionally, you can isolate the timeframe to see a granular pie chart that identifies activity types and allows you to drill down into individual workflows and expressions.

Workload activity chart is shown here:

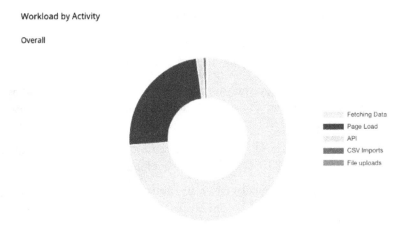

Figure 12.2 – Workload by activity (Please refer to the ebook version for colored references)

You can visualize workload usage by specific categories; this chart can help you identify what exactly is consuming more WUs and so help you focus on improving or reducing the usage in specific areas of your application. This chart is interactive and lets you click to remove specific categories and have a better picture of what areas are consuming the most; you can also click on it to go to the specific area of your Bubble app that is responsible for that operation.

Can I view the real-time workload data for my app?

Yes, real-time workload reporting allows you to view and access workload data in real time. The bar chart provides one-minute granularity for the last 24 hours.

How to optimize WUs?

To optimize your app to reduce workload consumption, you can do the following:

- Use the workload metrics dashboard to understand your app's current workload consumption and identify the processes consuming the most workload

- Prioritize workload optimization as part of your development goals, balancing it with user experience and functionality

- Optimize searches by setting them up efficiently, avoiding nested searches and advanced filters, and using as many constraints as possible

- Pay attention to what happens on the page load, as it's a key area for improving workload consumption

- Set up efficient workflows by questioning the necessity of each action, using lightweight conditions, and optimizing dynamic expressions within actions

- Optimize backend workflows by considering the workload spent on scheduling, conditions, and actions in API workflows

- Reduce the workload on database trigger events by assessing how often the associated data type changes and combining multiple triggers when possible

- Make strategic decisions about whether to use database trigger events or make changes directly in the original workflow

- Always weigh the importance of server-side processes for security and database integrity against the goal of reducing workload

Remember to monitor the workload dashboard to help identify key areas for optimization, and keep in mind that not all optimizations may yield significant workload reductions. For sure, paying attention to the WUs is an important part of your application but not all. In the next section, we are going to continue learning more about the maintenance of your no-code apps.

App management – Versioning control, updates, and production to Live

When building a digital application, the work is never done. If you don't know that, you must understand that the deployment of an application doesn't mean it is done; a digital product is never done, finished, or completed. On the contrary, when you launch a product, it just means the journey has just begun. Luckily, working with digital products allows you to test things fast and fix issues by simply pushing a button. This flexibility is key to continuously improving products and finding out what works best. Since the dynamics of building products involve always improving them and making them better, it is important to keep track of your versions and updates, managing what is still in production and what is live.

Versioning control

Depending on how you or your team work together, there will be cycles of development; in the agile world, we call them sprints. They usually last one or two weeks of work, and during this time, you will be creating new features or improving the existing ones. After that period, a new version of your application should be ready to be sent for production. When you work on a setup like this, **version control** becomes an essential part of the process so you can manage changes that are being made to your application and have a backup of versions in case you need to roll back. Version control allows teams to work on independent versions simultaneously. Remember that there is a difference between what is on the development side and what is in the live environment. When working with version control, you will always be choosing what is inside the "**Main**" (development) and "**Live**" (public-facing) areas; imagine them as two different worlds. If you are familiar with Git, this concept might be already familiar.

Bubble has a version control feature that allows you to keep historical records of alterations, enabling you to track, revert, or merge modifications made throughout the app. The basic features are available on any paid plan, and the advanced ones are available on higher-tier plans; the free plans don't have this feature, as it is not possible to deploy an app if you don't upgrade to a paid plan. In basic version control, Main is the sole branch for development changes and deployments. Premium version control introduces advanced branching capabilities, allowing **custom branches** in Main, **hotfix** branches from Live, and **flexible merging**.

What is a branch?

A branch is a parallel line of development that allows developers to work on different aspects of a project simultaneously without affecting the main or primary codebase. Imagine this: there is a development version and a live version of your app, and anytime someone from your team or you want to create a new feature or change something, you can create a new branch, a separate environment to work on independently, and another developer can do the same and create their own branch. When everyone is done with all the changes, you can merge things before the deployment. If you both changed similar things, there will be a conflict that you will have to solve, choosing which changes in that branch should be considered and which ones should be disregarded. Once the merger is done, both parties bring their changes to the Main environment and can then deploy the changes to the Live environment. Imagine this process as a train track where you create sidetracks to work on changes and, at some point, bring everything back to the main track. Working this way allows multiple people to work together without having to work simultaneously in the same environment. It also allows you to break down changes and change smaller parts of the application in separate batches, which is another key concept when working in a lean and agile environment. By using versions whenever you deploy something new to the Live environment, you can control your app's integrity, facilitate collaboration among your team members, and experience more revert bugs that may arise from changes. It also helps you track potential conflicts between versions, especially when you are working with branches.

Updates and production to Live

This section elaborates on best practices for managing updates while preserving version control. It emphasizes the importance of maintaining consistency across versions, implementing versioning strategies, and establishing protocols for seamless update deployment. By adhering to meticulous procedures, developers can ensure a systematic approach to updates, minimizing disruptions and ensuring a smooth user experience.

Updates are important to help you keep the application current. When you discover a problem that must be fixed, need to obtain user feedback, need to think of a new feature that must be developed, or realize your app is falling behind the competitors, an update will be needed. It is a natural process of developing apps; they will need to be updated from time to time; otherwise, they will get old and may become obsolete. It is better to adopt a regular and proactive schedule and keep updating your app than let it die slowly and only realize it too late. You don't have to do updates every day or week; you can define when to do them. However, updates are going to be required from time to time.

For instance, an update can be a plugin that needs to be upgraded to a new version.

Sometimes, it will be something that can be improved or changed. Maybe your home page needs a new, fresh look because the users are getting bored with the same visual.

There are a lot of reasons why you should update your application, but the most important thing is for you to define what the update cycles will be and what will drive these changes. It is also a good idea to align it with your business goals.

For instance, companies that are building startups usually have a long list of features they want to develop. This is usually a backlog or a product roadmap.

However, the reality is that they don't develop all of the features at once. At the beginning of the project, they will first build an MVP with just a few features.

Then, after this MVP is validated, they will keep adding more features as they go. As the startup learns from its users and understands what is working and what is not, it continue building new features. These new features will be their updates. Depending on the size of the startup and the goals, these updates can happen with a pre-defined cycle, which is what we usually call sprints in the agile world.

A good idea is to base your updates on user feedback and what you have learned during the launch of each version. Take the time to analyze the data you collect between each version and base the updates on the features and resources your users are asking for or pointing to that they need the most. Implementing metrics and analytics tools can help with that.

Whenever you have a new update to publish to production, you will do a deploy, sending that new version to the live environment so your users can interact with it and get value out of this new updated version. In *Chapter 11*, we covered the step by step of how to do this. Remember to keep the development and live environments relatively close, making sure that the databases are not too different so that you have a good reference to test your app before you do the deployment. Note that if you choose to host your data on a separate environment that is not inside Bubble, for instance, using services such as Firebase, Airtable, or Xano, this can also change your updates workflow.

If your database is not inside Bubble, you won't need to worry about managing the development and live databases but only the API integrations used to power your app with external data. Remember to adapt your process to what makes more sense with the setup you choose and how you work inside Bubble.

Updates can be done without communicating changes to users if you are, for instance, fixing bugs or making small improvements. However, if you are changing things that have a great impact on the product and deliver new features that users wish for, it is a great idea to communicate with them about these upgrades. A good practice is to send an email message to the users telling them about the new features and updates made to the product; this can also help with user engagement and make people want to use your app even more. You can also create a page dedicated to sharing updates with your users so that they can go there to learn about what is new. Another good idea can be to create live events or videos, showing people what is new and how to access and use the new resources. Remember, people are using your app to solve a problem. If you are building new things and improving a product that will help them, this is a good thing, and it is news they care about and will be happy about, so tell them about it. Tell the world about it; this can help your app grow in many ways, so don't be shy and spread the word about any updates and improvements to your product. In the next section, we are going to talk about app governance and how to handle applications inside your company after they have been developed.

App governance – Defining responsible people in your organization

App governance becomes an important topic when a new application becomes part of your company. If you are building a startup, it is natural that the CEO and development team will be very focused on the app and making sure it is being taken care of, but if you are using no-code tools to create solutions for a company that primarily has other types of applications or core activities, for instance, during a digital transformation process, it will be important to define who is going to be responsible for maintaining the application and the future development, maintenance, and decision-making processes.

With a culture of citizen developers emerging inside companies, which is a good movement, more and more apps are going to be created by different individuals in different areas of a company. The impact is positive but also has to be handled in an organized way. The bigger the company is, the greater the number of created apps will be, and together, these apps will represent a considerable piece of the company's operation.

The app governance process consists of defining the roles and responsibilities of stakeholders involved in app development, including key responsible individuals who ensure the company's guidelines are being followed. You can choose to create a centralized process where new apps have to go through an approval process and create a new area inside the company to review and maintain applications after they were created.

Another option is to create a decentralized process that follows a set of common guidelines to ensure quality and standard rules based on the company content. The most important part of this process is to define key responsibilities and ensure effective communication. After an app is created, it has to be clear who is going to be responsible for keeping it alive and running the maintenance and updates. You can define that the person who created it will do this, or after the creation of the app, it will be incorporated into the IT area or another area that takes care of your digital applications. You could even decide to create a no-code department inside your company just to handle new applications and, in parallel, train your company employees to allow them to build their own apps. Once apps are created by citizen developers, you can define what happens next, e.g., if the apps are internalized by the no-code department or if the app owner will be responsible for the app until they change their position or quit the company.

If the app creator leaves the company, make sure the app will be transferred to someone else or the no-code department; a good practice is to ask the app creators to build simple documentation to help others keep that information inside the company. If the apps being created are core apps and are essential to the business operation, it is advisable that they move into a department with more resources that can take care of it. For smaller and simpler apps, you can give individuals inside your company more freedom to create and use them based on their daily activities and needs.

Implementing no-code tools inside your company and allowing people to become citizen developers can be an empowering movement that can take your company to the next level by democratizing the usage of technology. No-code tools are truly empowering the next generation of builders across the globe. So far in this chapter, we talked about the importance of controlling versions, updates, and maintenance. In the next part, we are going to learn a bit more about analytics tools and ways to monitor our application, which form an essential part of ensuring we understand what is going on with our application and our users.

Monitoring app performance and user analytics

In the previous chapters, we talked about the various tools you can use to test your application's performance. These tools are very helpful for you to test things before your app's deployment, but they are also important in the long run. Some tools should be manually checked from time to time. It can be always before a new deployment or periodically; you decide. Other tools allow you to set up alerts and let you know in case things go wrong. As mentioned before, there are a lot of tools you can use right inside Bubble to help you monitor your app via the existing dashboards, but you can also plug in new tools on top of your application that can send you messages and notifications to let you know if the application is working properly or not. We recommend you check the plugins marketplace to find helpful plugins. For instance, you can track if your domain is up, if the app is working, and how your users are interacting with your product. You can have tools that are used internally but also use others that can help your users know if your application is working properly; a common example is building a status page that tells your users if your application is up and running, which can also reduce support messages and angry customers in case your application goes down in an unexpected event.

Make sure to set up monitoring tools based on what you think is important to measure and monitor, but also be careful not to track too much and end up in a monitoring nightmare. The more your app grows and the more important it becomes for your company, the more you will want to track, and that makes sense, but make sure to prioritize what is the most important first and grow as you obtain more resources to dedicate to monitoring. When your company grows, and you have a development team, this will potentially become part of a routine made by your team or a specific department, but at the beginning, it will be just you, and having a couple of systems in place will be enough. Having no monitoring tools and routine can be a mistake. Monitoring too many things can be a mistake, too. Just make sure to seek a balance. Only monitor what is truly essential and actionable. In the beginning, it is important to have a recurring schedule and to understand how you need to act when things go down.

User analytics is also an important piece when talking about monitoring, e.g., knowing what is going on with the app that you built and if users are interacting with it and using it. Set up at least one analytics tool to understand user behavior, such as Google Analytics or Hotjar. These tools track user interactions and allow you to quickly identify if things are going well or not. You can define alerts or even create reports, but one of the most efficient ways to monitor user behavior is to define a routine for logging into the analytics tools and checking the numbers. For instance, you can define a specific day of the week or month to check how the users interact with your app. You can create cohorts and analyze the metrics to know if users are converting well and if they are returning and recommending your tool to others. Remember, part of the monitoring is knowing if your app is running and stable infrastructure-wise. However, when thinking about building a product that solves a problem (especially for startups), user monitoring and analytics can be as, if not more, important than having a working app; after all, your app can be live and operational, but if no one is using it and getting value out of it, then you still have a big problem to solve.

Handling maintenance and addressing user feedback

Effective app maintenance involves an ongoing process to address bugs, optimize performance, add new features, and incorporate user feedback. Responding to user feedback, whether through direct communication or feedback mechanisms within the app, fosters user engagement and loyalty. Having direct contact with your users is also very beneficial in helping you understand if you are building the right thing. Combine user feedback with analytics tools so that you have qualitative and quantitative data to evaluate when making decisions.

Embracing user feedback is important to keep your app growing and better satisfy user needs. However, this also doesn't mean that you are going to build everything users ask for; be mindful about what you will and won't build based on user requests, and remember that you, as the founder, have to know your end goal and keep your vision alive while guiding the development of the app. Do listen to the users and always collect user feedback; that is important, but also know when to implement what they are asking for and when to completely ignore it. To collect user feedback, make sure to create direct channels and easy ways for your users to communicate with you. A few simple examples might include adding a feedback button on your website, adding a link to a form, or actively sending emails with surveys, asking for feedback. You can also choose to use tools such as Hotjar that will help you do that without worrying about developing such features.

In an ideal scenario, you will be shipping new features and testing new things regularly, especially if you are a startup and still figuring out product market fit. If changes are small, you can ship things during the day and users won't notice the impact.

There will be times when the changes to your application are too big, and it won't be possible to just deploy a new version of your application without anyone noticing such a big change. As mentioned before in this chapter, announcing changes and letting users know what is going to happen can be positive and beneficial, but sometimes, we know users might not like it if the application stops working for a few hours or days.

In cases where you need to run a bigger maintenance process, a good practice is to find a time of the day (or night) when users are less active so that the impact is lower; you don't want to put your app into maintenance mode during a time when most people are using it.

Another good idea is to tell people about your scheduled maintenance upfront and tell them about it a few days before it happens. You can send them an email or add some information right inside the application so that when they log in, they will see a notification about it. During the maintenance period, you can either lock the pages of your app with a message or use a plugin to help you do that. Under the plugins marketplace, you can find a plugin called **Down for Maintenance by Zeroqode**. This is just one option; if you search for a similar term, you can find other options as well.

> **Important note**
> Remember, your application is hosted by Bubble; if their servers are down or are experiencing maintenance, it is important that you also replicate the message to your users since it is going to impact your app as well.

Handling maintenance and user feedback is something that will follow your app development routine as you go, especially after the first official launch. A lot of first-time entrepreneurs think that the deployment of an application is like the finish line of a race; once they get there, it is over. That is far from the reality. Imagine the launch of an application as just the first milestone of an infinite journey. Well, it can be finite if the startup dies. However, if you persevere, keep in mind that maintenance, improvements, and user feedback loops will keep going as you grow and that this is a good thing; it will allow you to fix problems as you go, improve the application based on feedback, and help people solve their problems in a better way.

Summary

In this chapter, we've covered the essential aspects of managing your app by focusing on monitoring, maintenance, and updates. After publishing your app, these topics are crucial for a successful web application. You've learned about dashboard metrics, which provide insights into app performance, and we have delved into workload units (WUs) for measuring resource use. We've also discussed app management, covering updates, version control, and the different environments inside Bubble, production, and Live.

This chapter also guided you on how to handle your app's governance and ongoing maintenance, considering user feedback and analytics tools to deliver the most important features to your users.

Understanding these concepts is important for maintaining a live app and continuously building a strong and user-friendly experience, ensuring sustained success while prioritizing user feedback. In the upcoming chapter, we will dive into advanced Bubble techniques to help you optimize the performance and scalability of your application.

Further reading

- For more information about plans and WUs, visit the Bubble manual: `https://manual.bubble.io/account-and-marketplace/account-and-billing/pricing-plans#workload-units`.

- For more information on version control, please refer to the Bubble official documentation available here:

  `https://manual.bubble.io/help-guides/getting-started/navigating-the-bubble-editor/deploying-your-app`.

13

Optimizing Performance and Scalability

In this chapter, you will learn about performance and optimization and how to improve your application as your user base starts to grow. You will learn a variety of tips on how to improve your app's overall performance, scalability, and the **user experience** (**UX**). We will cover some of the best practices and provide advice for efficient app design, performance, and scalability, covering strategies to handle databases and workflows and creating interfaces that resonate with users and provide a pleasant experience that also impacts the perceived performance of your application.

This chapter is a toolkit to optimize your Bubble applications for superior performance, scalability, and user satisfaction. Let's dive in!

By the end of this chapter, you'll have gained a comprehensive understanding of optimization methods and how to improve your application performance to allow it to scale and grow.

In this chapter, we're going to cover the following topics:

- Why is performance important?
- UX and **user interface** (**UI**) tips – best practices for your app design
- Efficient database usage – improving workflows and queries
- How to minimize page load times
- Is it possible to scale your Bubble app? – Real examples
- Scaling your Bubble app for increased traffic
- External databases as an alternative option
- Basic tips to improve your app's overall performance

Why is performance important?

It might sound obvious that performance is important, but why exactly? Performance is crucial to app development as it directly impacts user satisfaction and engagement. No one wants to wait for a page to load or spend too much time figuring out how something works; it must be fast and intuitive. A well-performing application ensures seamless UXs, fast response times, and efficient use of your app. Remember – your app should be a solution to someone else's problem, not another problem. Moreover, performance optimization contributes to enhanced reliability and scalability, allowing your app to grow and accommodate increasing user demands without sacrificing functionality or speed. A strong foundation is important to allow you to grow as you go.

Key points to consider for app performance include the following:

- Fast response times improve user satisfaction and engagement

- Seamless UXs lead to higher retention rates

- Efficient use of resources ensures optimal performance and scalability

- Reliability is essential for maintaining user trust and loyalty

- Scalability allows your app to accommodate growing user demands without compromising performance

- A well-optimized app foundation enables smooth growth and evolution over time

Translating performance to the users' eyes means how responsive the application behaves when users interact with it. How do they perceive it? Performance is a complex combination of hardware, software, design, and human interaction, it depends on the servers that are hosting the application, the code that is being executed behind the curtains, the device a person is using and its capacity, the browser requesting and reading files, the internet speed, the overall UX and context of a user who is, at the end of the day, trying to get something accomplished to solve a problem.

The performance of your application can be impacted by various factors, and little things can help to make or break your app's performance, such as the following:

- How long your website and pages take to load

- How responsive a button is after it is clicked

- How fast new elements on a page take to appear

- How well displayed and smooth animations are

- How easy it is to navigate from one page to another without loading

- The time it takes to load new data or display information on a page

These are just a few examples that count toward the overall performance experience; as you can see, not all of them are related to actual performance issues caused by the server or the code itself. There are actual performance scenarios and user-perceived performance; a lot of it can be related to the UX and simply communicating actions and diminishing the perceived loading time to make the overall experience a little smoother. In the next section, we are going to talk a little bit more about UX and UI tips to help with your app's performance and overall UX.

UX and UI tips – best practices for your app design

UX and **UI** are integral aspects of app design that significantly influence user engagement and retention. If you are not a design expert, the most important thing is to not reinvent the wheel. Follow existing design patterns, and don't get too creative in this area; remember that people are used to using the web already. Build things that are already familiar and like other existing websites so that the users don't have to learn how your app works entirely from scratch. If you can, hire an expert to help you; a UX/UI designer can help you a lot in this process. If you can't hire someone, at least try to simplify the app idea, make the design simpler, or use a good and well-designed premade template; be careful with the bad ones as well. When using templates, also make sure you don't break them or make them look bad. Again, search for references and best practices; we will cover some of the basics in this chapter.

When building your app, remember to create intuitive navigation. Define clear page names and an easy-to-understand visual interface. The navigation flow should be intuitive and simply work. Structure your layout in an organized way, always thinking about the main elements needed on that page so that the user can achieve their goal, which will help you enhance usability. Less is more!

Remember to prioritize the user and their goals and needs; your application has to be simple and help them accomplish tasks. Prioritize user-centric design principles, such as clear communication, consistency, and responsiveness. Design and great UX come with time and practice, but one key step to achieving better usability is through testing. Do not spend too much time building new features without testing them with real users; this process is key to finding potential areas of improvement. Apps that are now successful and have a great UX went through numerous iterations and improvement cycles with real users to get to where they are right now, and even now, this process isn't over. So, keep your development process aligned with this mindset of building and testing with users to learn what could be improved as you go.

UX and UI best practices and tips

You might not be a designer by trade, but still, it is very important to pay attention to the UX and UI of your application. People like to use software that is easy to use and understand, and usually get really angry and frustrated when it isn't, so to avoid common mistakes and increase the chances of building something users love, here are a few tips and best practices to help you:

- **Keep it simple**: This one is obvious but commonly neglected. It is easier to build something more complex than it is to build something simple that works. As much as possible, try to ask what can be reduced and simplified. Simple interfaces minimize cognitive load and make navigation intuitive. The less you have to explain how the app works, the better and easier it will be for people to just jump in and use it.

- **Consistency is key**: Build a system and repeat it everywhere. Use design elements such as colors, typography, and layout to create a cohesive UX. Do not reinvent the wheel or change styles and patterns too much; stick with what works and make your application familiar everywhere. Once a user learns what one component and element does, it will be simpler to understand how the whole system works because of the consistency; not only that, but it will also help your branding and visual identity to be more recognizable. Using or building a design system can be a great thing to consider.

- **Put yourself in the users' shoes**: Empathy is very important when designing anything. While seemingly straightforward, this principle is crucial and sometimes quite hard. To achieve an optimal UX in your website or app, understanding your users is imperative. To build a good solution, you must understand the users' needs and emotions. Utilizing a customer journey map can effectively help you visualize the potential experiences of your target audience and areas of improvement.

- **Optimize for mobile**: If your users will use your application on a mobile device, you can have a specific mobile version, or at least make sure it is responsive and provides a seamless experience across different devices and screen sizes.

- **Prioritize content**: Content is king; make sure people can get the information they need easily and provide ways to guide users through the app's core functionality.

- **Build a sitemap**: It is essential to provide a clear navigation on your website. A sitemap shows how pages are organized and connected and how the user flow might happen. Organize your application sitemap in a way you and your users can easily find information and navigate. When you add new pages, refer to the sitemap to decide where they should go. If your project has too many pages – let's say, more than 10 – it may be a good idea to break them into small groups by using categories. For instance, you can create a category name to use as a label – let's say, **Resources** – and inside this category have multiple pages under it. Do the same for each group, and you will have a better and more organized project structure. Remember – when you delete a page, it becomes a 404 error. Keep your sitemap updated to manage these changes effectively and avoid users getting lost with broken links.

- **Clear navigation**: Make sure people can find the primary navigation of your application by using descriptive labels for menus and buttons to help users easily find what they're looking for. You can test how users understand a word or category to understand if it is actually conveying the right message.

- **Visual hierarchy**: Every element on a page plays a role; their size, shape, position, and color can alter how users interact with it. Hierarchy is important to highlight a certain feature or area on a page; by using visual cues such as size, color, and placement, you can emphasize important elements and guide users' attention to what is more important.

- **Feedback and confirmation**: When users perform an action, they expect something to happen. Providing feedback for user actions is very important, so make sure to create and pay attention to these messages. They may seem harmless, but they play a huge deal in terms of how the users interact with the application. Make sure whenever users perform an action there's clear feedback of what was the result of that action, and give them a path to proceed to the next step.

- **Accessibility**: Consider accessibility to ensure that all users, including those with disabilities, can easily access and use the app. There are plenty of tools that can help you spot areas that can be improved, and with simple actions, you can make it much more accessible, even if your application is just in the initial phase.

- **User testing**: As mentioned before, usability tests with real users can help you identify pain points and improve the overall UX. Don't be shy; just do it, and you will see how powerful and helpful it will be.

- **Continuous improvement**: Your app will never be done; that is just the reality – face it. Regularly gather user feedback and iterate on the design; there will be always something to improve. The key here is to discover improvement points but to also prioritize and focus on ones that can move the business forward faster. This, if turned into a recurring process, can for sure optimize the UX over time.

Here are a few more tips and best practices to consider:

- Differentiate buttons and links and create a primary and secondary button style. Make sure to label buttons correctly with verbs and clear actions; do not make them too long or confusing. You can also add icons close to links or inside buttons; make sure you use an icon that is related to the word or action and that is understandable by the users. An icon is not just something to make the app pretty; it has to help convey a message.

- Make use of colors to identify specific states and messages – for instance, when deleting an item, use the color red; for an alert, the color yellow; and green to indicate a success message.

- When creating forms, make sure the labels are well placed and close to the element. Avoid adding labels only inside the inputs so that once users start typing, they can still remember what that field is about. If possible, provide tips and extra information to make it clear to the users how to fill it out.

- Be mindful of error messages; make sure they are clear and that they indicate what to do next. These little things can build frustration and destroy the experience; imagine getting an error with a code and there is nothing you can do about it. It is just terrible. If you are not sure how to write good error messages, take a look at a few topics such as UX writing to get a few examples of how to write better copy on your application's interface.

- When adding popups, make sure users have a way out and that the **Close** button or icon is visible and easy to find. Hiding essential features and forcing users to go to a certain path is also known as a bad practice and is called "dark patterns;" you can look that up if you wish to know more.

- Make the page navigation easy to find and clear, and use titles on the pages to quickly help users identify where they are. A still useful technique is to use breadcrumbs. Don't forget about color contrast and readability; don't use too many colors and different fonts, make sure the background has a good contrast with the font color applied at the top, choose an easily readable font, and apply it with a decent size. If needed, there are tools online that can help you check if the contrast is good. There are also some tools listed in this book that can help you check how accessible your website is so that you can spot areas of improvement and make it better.

UX and UI design is very important and can require an expert in the field with years of experience to master. You don't need to be a pro; if you just take care of the basics, it will give you great results, and it is way better than not worrying about it at all. If you want to dive deeper into this subject, there are plenty of good books you can read, added in the *Further reading* section.

Improving the UX and UI of your application is just one part of the equation. Now, let's dive deeper into how to improve databases, workflows, and other aspects of your application that can also help with scalability and performance.

Efficient database usage – improving workflows and queries

In this section, you will learn how to improve database queries and workflows to improve your app's performance. Part of this process is understanding the key concepts involved and the mindset, and part of it is implementing a few specific practical techniques to reduce the app loading time and improve performance when using databases and workflows inside your Bubble application.

The mindset

Now, let's start with the mindset part. The key here is to make it simple; of course, this is obvious advice, but trust me – it is very easy to overcomplicate and overengineer an application; your goal will be to focus on how to create things in a simple and minimized way. When talking about workflows, how can you reduce the number of steps? How can you reuse similar actions in different parts of the application? How can you make it less complex? When talking about databases, how can you make the database smaller, with fewer fields and less information? Can you break it down into smaller pieces? For databases, you have two goals: reduce complexity and make them lightweight. You can think of databases as little packages of information that will be moved from the inventory to the store. Imagine every time someone is loading a page, they are asking someone to get a package from the inventory and put it on the storefront: this is your database query, specifying which package and what information should be moved from one place to another. So, the idea here is simple: the more packages you ask for and the heavier they are, the more work will be required to move them. It is like a download. So, to improve performance, you need to control requests and make sure that the file that is being downloaded is as light as possible. The lighter and less complex it is, the faster it will be to retrieve it.

Databases

OK – you got the idea. Now, let's get more practical into techniques to improve workflows and databases. Let's first talk about databases. When you're retrieving records from a database – for instance, the `User` data type – it may include fields such as first name, last name, email, and so on. In that specific scenario, the download size for that database request would be relatively small. However, if you're fetching the same database but with a few more fields such as the user address, phone number, profile picture, payment information, parent information, related field, and so on, the amount of data will significantly increase, and so the download of that set of information will as well. Imagine a database for blog posts, each containing a lot of fields and a heavy amount of content inside each field; the download size can become larger, especially when dealing with a large number of records. This accumulation of data, even in seemingly small amounts per record, can result in a significant increase in kilobytes or megabytes to be downloaded when querying a database. Now, multiply a single item on the database that may have just a few bytes when querying multiple items. It may become a significant number added to the page components, images, and everything else that has to be loaded to render your application.

So, what is the main takeaway here? Minimize the number of fields and data types on your application, adding only what is truly essential. If you must add more fields, try to break them into other database columns and use relationships. This technique is known as satellite databases, which can help you reduce the initial loading time of a query and only ask for more data when needed using relationships.

Another method to improve your queries is by adding constraints to a search, instructing the platform to download only the necessary data. Note that Bubble will download all fields of a data type, even if they are not displayed on the screen.

An alternative is to use option sets instead of databases. Option sets in Bubble function as a lightweight database for your application. The main difference is that you can't modify option sets without redeploying your app. However, they have their advantages. Option sets do not depend on your database and are actually embedded in the Bubble code downloaded to the browser. This means you don't need to query the Bubble database to access their information, meaning this can be an alternate option in some specific cases, reducing the need to query more databases. Option sets serve as a fundamental tool for establishing static values within your Bubble app. For instance, they're commonly used to create categories. Since these values rarely change, the need to redeploy the app for updates isn't a concern. However, it's important to recognize that option sets aren't intended to store large databases. Remember that your application should only download the necessary information. Full option sets are downloaded each time a user accesses any page, so use them responsibly.

Workflows

Workflows also play a huge part in the experience of an application because they are related to user interaction. When users click a button, they expect things to happen and to happen fast. That is why it is important to understand not only how to improve workflows but also how to manage users' expectations and communicate things when they happen.

In previous chapters, we talked about workflows; they consist of a series of actions that occur in sequence based on specific inputs, scheduled times, or conditional triggers. The performance optimization of a workflow depends on various factors such as its functionality, execution method, timing, and conditions. Workflows contribute to the overall data downloaded and stored in the user's device memory, impacting UX directly. Workflows are highly visible indicators of app performance as users can easily perceive the time taken from initiation to completion of an action.

We can divide workflows into two categories: frontend workflows and backend workflows:

- Frontend workflows are triggered and executed on a page level. They are actions performed and activated by page loads, user actions, periodic intervals, or specific conditions. Their performance can vary from instantaneous to visible delays during user interactions.

- Backend workflows are managed in the backend workflow editor and are not tied to a specific page; they are global to your entire application. They can be triggered or scheduled from any page in the app, making them globally accessible. These workflows can be initiated immediately or scheduled for future execution through frontend actions, external service connections via APIs, backend triggers, regular intervals, processing lists of items, or self-scheduling (recursive workflows).

A good piece of advice is to understand the difference between client-side and server-side actions; this concept can help optimize workflows as you can consciously choose when to use one or the other. For example, let's consider a scenario where you have an e-commerce website built on Bubble. When a user adds an item to their shopping cart, you want to provide immediate feedback by updating the cart icon in the navigation bar to reflect the new item count. Here's how you might approach this using both client-side and server-side actions:

Client-side action: The user clicks the "Add to Cart" button.

Client-side action: Bubble immediately updates the cart icon's display to show the incremented item count without waiting for server confirmation. This action, such as setting a state or changing an element's visibility, occurs instantly on the user's device.

Server-side action: Similar to the previous scenario, the user clicks "Add to Cart."

Client-side action (initiated): The client-side action triggers a server-side workflow to update the database by adding the selected item to the user's shopping cart.

Server-side action (execution): The backend workflow executes, processing the request to add the item to the cart in the database.

Server response: Once the server completes the action, it sends a response back to the client.

Client-side update: Upon receiving the response, the client updates the cart icon to reflect the updated item count.

In this example, the client-side action provides immediate visual feedback to the user, enhancing the perceived speed of the application (UX). Meanwhile, the server-side action ensures data integrity by updating the database accurately. By leveraging both client-side and server-side actions effectively, you can create a seamless and responsive UX while maintaining robust backend functionality.

The key here is to inform users about changes and create responsive actions that let the users know their operation is in progress. If there is no way to make it faster, at least communicate and give prompt statuses to the users so that they at least feel that things are working well and in progress. Workflows can take longer, especially when dealing with database operations. For example, if you're displaying a list of items in a repeating group and then executing a workflow to change one specific data to another (let's say it is a list of cars, and you want to change all the cars on the list to a specific brand), Bubble can execute this action almost instantly because it already has all the necessary data downloaded. However, if you try to execute multiple searches and add filters and variables to the workflow and database query, this becomes a more complicated task, and delays can occur when introducing more variables into the workflow. In that case, you may choose to use another method of running that request, maybe running that in the background and informing the users that this will take time, and either show a loading status or inform the user later when the operation is done, via the application or via an email message. It is always a balance between technical improvement and managing users' expectations. Just as clear communication is vital in human relationships, it's equally essential in app development. Imagine planning a road trip with friends; updating them about route changes and potential delays can ensure everyone remains informed and engaged during the journey. Similarly, transparent communication in app design fosters trust and engagement among users during their navigation path.

For instance, imagine a navigation app guiding you through a city. Visual cues such as street signs and directional arrows keep you informed and reassured about your journey's progress. Likewise, in app workflows, visual progress indicators serve to acknowledge user actions, indicate ongoing processes, and signal task completion, ensuring users stay informed and engaged throughout their interaction.

Imagine you're using a food delivery app. After selecting your meal and proceeding to the checkout, you click the **Place Order** button. Immediately, the button changes color to indicate that your action was performed. A loading animation appears, letting you know that your order is being processed. Finally, a pop-up message confirms that your order has been successfully placed and provides an estimated delivery time. Throughout this process, the app communicates effectively with you, ensuring a smooth and transparent UX. Sometimes, it is not always about the technical part but also the visual elements and overall design of your application that can convey trust and seem performant.

How to minimize page load times

Minimizing page load time is one of the ways to improve your application performance, and there are a lot of little things you can do to make sure your page loads faster. When your page loads faster, your user will perceive your application to be more responsive and smoother. This sensation gives the user the understanding that your application is performing well. Note that you can also use techniques to improve the perceived loading time, which in some cases doesn't necessarily mean your application is loading faster in reality, but the users will think it is loading faster due to the strategies applied, and in some scenarios, that can be enough to deliver a seamless or better UX while using your application.

Think of page loading time as a telephone call. Your web application is running on the user's browser, which is their telephone. When they type in your website URL, they are calling the server – in our case, Bubble – and the host that has all the files and information that make your application run. During this call, there is time between the files in the host getting to the user's browser to answer the call. The more files your application has to download, the more time it will take for that call to be completed, and therefore the more the users will wait in line. Websites work with an HTTP request call, which means two computers communicating via the internet, one trying to get information from the other one. It is like making a download, but through your browser, to render a page and to run an application. The more distant the server is, the more files there are, and the bigger they are to be downloaded and rendered, the more your application page will take, so the idea here is to always simplify your application page and file sizes so that this call happens faster.

Now, let's look at some simple techniques you can implement to ensure your page load time can be minimized:

- **Optimize and compress images to reduce file sizes**: When your application loads, it will download every image added to your pages. If you don't necessarily need images on your page, simply avoid adding them; this will for sure increase page load time. If you do need to use images, make sure you compress and optimize them using services such as `tinypng.com` that can reduce the file size without losing quality. You could also search for plugins that can help with that. Another tip is to make sure the image size is accurate for the type of usage you will have. For instance, do not add an image to the page that is three times the size it will actually be displayed at; it will only take more time to load and make your page slower. Also, consider the image format; there are many options, such as `.JPG`, `.PNG`, or even vector formats such as `.SVG`. It will depend if you need a transparent background or not, but usually, .PNG is one of the lightest options and most used formats. If your image is a vector, such as a logo or icon, you can use .SVG and compress it; this can make your logo expand to any size but will still be very low in terms of file size. When thinking about images, icons should also be considered. Don't add too many, and if you do, make sure they are either .PNG with an optimized file size or even .SVG or a lightweight font instead of using images.

- **Reduce CSS, JavaScript, and HTML**: Bubble will do some part of the work, but there are some aspects of Bubble you won't be able to change; that is part of the core of the application that only Bubble developers can handle. But you can control how much you add to your pages, so be careful and conscious about the extra code you add, meaning CSS, JavaScript, and HTML. When you add custom plugins, they will add more code to your pages and require it to load when the page is rendered, so make sure that you only add what you need. If you build a plugin yourself, minimize the CSS, JavaScript, and HTML to ensure it loads faster. The number of requests made is also important, so the lower the requests, the better. When building your pages, use similar styles and repeat the same styles for buttons, fonts, colors, and so on. This can also help with the amount of classes and styles generated inside your application CSS.

- **Use just a few fonts on your project**: When you choose fonts, they will also count toward page load time, as fonts will also be files that need to be requested and loaded to display the page. It is advised to stick to up to three fonts on your project, preferably one or two at most. This will help your project performance but also overall design; using too many fonts can be a bad idea for performance since it will increase the size and amount of files being loaded and also make your page layout look bad if you combine too many fonts, especially if they don't match in terms of style. Another tip is to use fonts to write things on a page rather than using images because the font itself will be rendered on the page way lighter and faster than loading a new image that has something written with the same font. So, if possible, instead of creating images that hold text for some reason, try to create them using actual page elements – for instance, if you have a banner, maybe creating that banner with actual components using, for instance, a font and a button that are part of the page can make more sense than creating an image that has the same style and characteristics.

- **Implement lazy loading for images and content that isn't immediately visible on the screen**: Lazy loading is a technique that delays the loading of non-critical resources, such as images or videos, until they are needed, improving initial page load time and reducing the amount of data transferred. You can do it manually by controlling when a component on a page will load or use a plugin to help you with that task.

- **Reduce the number of lists and database requests on a page**: If you request too many details about a record or load a list with various results, this can increase page load time, so whenever possible, query only necessary information and let users request more data or more items loaded on a list. This is also related to the lazy load method and will allow you to save resources from the initial loading time and let users request more data as needed – for instance, instead of showing a list with 100 items, load just the first 10 and let them click a button to load more or show the next 10 items.

- **Build multiple pages in one**: This can be a smart technique but has also to be used carefully. You can build a multi-page instead of a single page. The difference is that multiple page contents will be loaded on the same page and only displayed when users interact with certain elements. This can help load the page just once and have all the content available in a fraction of the time when users click a button to change to another page, for instance. This will be faster because the second page is already there, fully loaded; it was just not visible at first. Depending on the type of project and application, this technique can be very useful and allow users to have a great UX. On the other hand, you have to be careful to not build a bloated multi-page and end up having a too-slow initial loading time or adding too many components to the same page to the point where the browser can have trouble rendering it. Remember that users might be using different browsers and computers, and your page can become too heavy for their computers to run, which will have the complete opposite effect of the intended action, which was to provide a faster application to the users.

In summary, remove any unnecessary components, images, icons, fonts, files, content, and third-party scripts and plugins that may slow down page loading.

> Tip
>
> You can always refer back to the previous chapters where we talked about tools to help you test your application performance. Using these tools will help you identify areas of improvement and also check if the page loading time is already good enough.

As you may already have noticed, performance is very related to being reasonable and reducing unnecessary resources; it is about less being more. It is also about not repeating yourself, which leads to the DRY concept. The **DRY** (**Don't Repeat Yourself**) concept is a software development principle aimed at reducing the repetition of code within a system. The idea behind DRY is to promote code reusability, maintainability, and efficiency by ensuring that every piece of knowledge or logic in a system is expressed in a single, unambiguous way. With no-code tools, we are not writing code like a developer, but there are some principles we can also use and apply from the DRY concept. Instead of thinking about code, we can think about components and features we add to our projects when building it. To apply the DRY principle effectively, we can adapt the concept to our context and think about it like this:

- **Identify repetition**: Look for duplicates. Once repetition is identified, organize the application with reusable components, functions, or modules. Make sure you use similar components on your pages, reuse them, create a similar and consistent style and page structure, and use the same resources whenever possible instead of creating new ones. Design and develop with modularity in mind, breaking it down into small, cohesive modules that perform specific tasks. By following the DRY principle, you can avoid redundancy, improve the project quality, and enhance the overall maintainability and scalability of your projects. With simple techniques and the right mindset, you can for sure improve your page loading time; by reducing files and images and applying all the techniques and advice given in this chapter, your project pages will be lighter and so take less time to load, giving your users an awesome navigation experience.

Is it possible to scale your Bubble app? Real examples

The quick answer is yes, but at the same time, no. How is this possible? Well, an app that isn't built for scale simply won't scale. And sometimes that is OK; the plan was to validate something and get things done, not to scale. But if scalability is your goal, then yes – you can prioritize that and build it to scale. Accommodating a growing user base and increasing traffic requires careful planning and execution. Scaling a Bubble app, especially in the context of a start-up, should become a concern; otherwise, you might be building a scalable app that will never get there. To scale an app, you will have to keep improving it and making it perform well as the user base grows, as you decide to prioritize it. But it is possible. This is the reality with code or no-code. Trust me Bubble won't be your only concern when growing and scaling a company. Although it might seem a tricky road, it is possible; you just need to plan and move fast as your application grows and, if needed, look for experts in the field to help you with that growth.

In this section, you will discover a few successful stories and examples that can prove it is possible to grow a company and scale using no-code and, more specifically, Bubble.

Here are some key examples of successful apps that scaled and were created with Bubble:

- **Dividend Finance**: This company runs a solar panel financing platform for homeowners and a CRM for installers. They have raised more than $330m and processed over $1b of loans through a Bubble-built solution since 2014. Website: `https://www.dividendfinance.com/`

- **Cuure**: Provides personalized supplements. Their website and e-commerce store are fully built on Bubble. The company has raised €1.8m from top investors in Europe. Website: `https://cuure.com/`

- **Plato**: A mentorship platform for engineering leaders. Funded through YC W16, it raised $3.3m and has built its back office on Bubble. Website: `https://www.platohq.com/`

- **Comet**: This company, initially created with Bubble, scaled to $800k revenue before raising its first $13m round. Website: `https://www.comet.co/`

- **Teal**: Started with Bubble for their platform and raised $5m. The company helps people navigate and organize their career search, from job tracking to a content library of resume writing tips and interview training. Website: `https://www.tealhq.com/`

A great source for inspiration and study cases is the *Bubble App of the Day* section on the official Bubble website where they feature various apps and the stories behind them. To know more about it, visit the website: `https://bubble.io/blog/tag/app-of-the-day/`.

Another great place to discover successful stories is the Bubble *Showcase* page, available here: `https://bubble.io/showcase`.

As you can see, there are a few successful cases out there, some bigger and others a little more moderate, but they prove it is possible to get there. The reality is that the number of projects that succeed in the start-up game, unfortunately, is low compared to the total amount of projects born, but that isn't necessarily related to no-code tools or Bubble; it is more related to all the causes and risks involved in building a new start-up from the ground up. Tools such as Bubble and other no-code tools can be a real and viable option to allow entrepreneurs to build a successful business; scalability is just one of the many constraints in this journey.

Scaling your Bubble app for increased traffic

Each app and company will have a different process, a different application, and a specific scenario when they need to scale; sometimes, scalability doesn't necessarily mean only handling more users accessing your application but also hiring more people on your team. Most of the time, people are worried about the software but forget other areas of the company that also need to scale as the company grows, so have that in mind as well. Sometimes, your biggest concern won't be the tools or the software but hiring and training people, improving support, sales, customer success, onboarding, moving to a new location, creating new areas of your company, and so on. Many of these things can come before you actually need to invest more in software growth itself. Another common scenario is entrepreneurs worrying about scalability when they are in the initial early stage of their start-up, as if in a matter of days this would become a concern. The tough reality is that most apps won't become a success overnight; that is just how it is, so scalability won't be a concern in most cases for many months.

Now, when it comes to scaling your software and growing your application, regardless of your current stage, here are a few things you can consider and think about to allow your Bubble application to scale and grow:

- Run performance tests and identify areas that need improvement regularly. Reduce scope and unused features by understanding if features are being used with metrics and by talking to users. If not needed, you can remove them and reduce the complexity of your application.

- Consider using an external database if the internal Bubble database becomes a problem. There are a few options mentioned in this book. Offload files to an external database if your application handles massive amounts of files, such as videos, images, and so on. Consider using a specific service to host and serve these files to your application using integrations.

- Review the app structure and pages. Make sure all pages and links are working properly. Remove or group pages; if possible, organize pages under an organized structure. Create smaller, simple pages rather than big and complex ones for faster loading times.

- Fetch only necessary data on page load to improve performance. Pages fetching a smaller amount of data load faster than those fetching larger volumes. Fetching simple data types such as numbers is faster than fetching large datasets. Keep sorting and filtering as close to the original search as possible to enhance efficiency. Queries applying sorting or filtering directly at the database level are more efficient. Advanced filters applied with ":filter" are generally slower than those applied in the Search ("Do a search for"). Filters done "on the database" are faster than those applied after retrieving initial data.

- Review and optimize the app's data hierarchy and queries based on Bubble's optimization principles. Simplify query expressions whenever possible. Instead of making additional database calls, consider changing element states. Make use of behind-the-scenes scheduled workflows for expensive calculations to avoid performance issues on page load. Consider using "Schedule API Workflow on a list" for better performance with larger datasets.

- Prioritize scalability decisions when that becomes a real concern. Consider separating your website from your backend if you want to focus your Bubble application on the backend only. You could choose another no-code tool to build and maintain your website, separating the two environments and using subdomains for each application.

It is nice to note that Bubble automatically optimizes queries, resizes images, and caches JavaScript to improve performance.

Scaling a Bubble app for increased traffic involves implementing strategies to ensure optimal performance and reliability under high loads. This can be done on your application side, most of the time, and in other cases, it will be related to your Bubble plan and how much it can handle. But don't think that just upgrading the plans will solve all the problems if your application isn't built to scale. Techniques such as implementing a scalable database with tools mentioned in this chapter can help optimize workflows; queries and page load are all mechanisms that can help distribute traffic efficiently and prevent performance bottlenecks. Additionally, monitoring key performance metrics using the dashboard and logs panel can help identify potential issues and proactively address them to ensure seamless scalability. It is not an exact recipe; there are various components, and each plays a significant role in this equation. The key here is to implement changes, test, and continue the process to get to a stable version as you go. Now, it is important to notice that there are times when you will need to move to an external database; this is what we will talk about in the next segment of this chapter.

External databases as an alternative option

Bubble has an internal database you can use to build your applications; this means you don't necessarily need to use an external database in most cases. The internal database is very powerful and flexible, but it does have some limitations, which can be overcome when you use an external database. As a developer, it is important you know about the limitations and also alternative database options so that you have alternatives to choose from when you build your applications.

Bubble's database limitations

As with any tool or resource, Bubble also has its limitations. Even code has limitations; some might say no-code is limited as an excuse to not use it or favor other solutions, but the truth is that everything has limitations and we just have to deal with them, or at least know them to, when needed, overcome them and find solutions. Here are a few of the existing limitations you can be aware of inside Bubble:

- **Data type limits**: Bubble imposes limits on the number of data types and fields you can create within your app. These limits depend on your Bubble plan, with higher-tier plans offering more data type and field capacity.

- **Record limits**: There are also limits on the total number of records (entries) you can store in your Bubble app's database. These limits vary based on your plan, with higher-tier plans allowing for more records.

- **Performance**: As your app grows and accumulates more data, you may experience performance issues, such as slower response times for database queries. Optimizing your database structure and queries can help mitigate these issues.

- **Scalability**: While Bubble's internal database can handle a significant amount of data, it may not be suitable for extremely large-scale applications with millions of records. In such cases, you may need to consider alternative database solutions. We are going to explore these further.

- **Data privacy and security**: Bubble's internal database is hosted on Bubble's servers, which may raise concerns about data privacy and security, especially for sensitive or regulated data. Developers should ensure they comply with relevant data protection regulations and implement appropriate security measures to safeguard user data.

Integrating external databases offers an alternative solution for managing data in Bubble apps, providing flexibility and scalability beyond the platform's native capabilities. By connecting to external databases, you as a developer can leverage advanced features and scale storage capacity to accommodate large datasets.

It is also important to point out that using an external database should be a conscious decision because in most cases, using the internal database is going to be a viable option and the simplest way to build your app, so don't rush into using an external database unless you know what you are doing or have clear reasons to do so.

When using an external database, your app complexity will increase, and you will need to be able to handle the usage of an external database; it will add another service to the bill, require you to manage and use two different tools, and require you to understand how to use APIs and the database service and technology you choose to use. Now, if you know you need an external database and are ready to use it, here are a few options you can consider:

- **Xano**: This is one of the most used and powerful backend development platforms used by no-coders. It allows developers to build and scale applications without managing infrastructure. It offers features such as database management, API creation, and business logic automation, enabling rapid development of full-stack applications. It is possible to integrate Bubble and Xano and use it as a database for your application; you can do it via an API or use an existing plugin available inside the plugins marketplace to make things easier for you. The plugin is available here: `https://bubble.io/plugin/xano-connector-1667758971605x316889048071536640`

URL: `https://www.xano.com`

- **Backendless**: This is another platform that provides developers with a set of tools and services to build and deploy server-side applications without managing servers or infrastructure. It can be used as a backend database alternative for your Bubble app. To implement it, you will use the API Connector; there is a tutorial available here: `https://backendless.com/how-to-integrate-bubble-io-with-backendless/`.

URL: `https://backendless.com`

- **Firebase**: You can also use Firebase, which is developed by Google. It offers a wide range of services including real-time database, authentication, hosting, cloud functions, and **machine learning** (**ML**). Firebase enables you to build a database in a very easy way, and it can be integrated with your Bubble application. One of the main reasons to use Firebase as a database alternative is the price. You can do the integration manually using APIs or use a plugin to help you with the task. Here is one plugin that can be helpful – Firebase Realtime Database Plugin for Bubble: `https://zeroqode.com/plugin/firebase-realtime-database-plugin-for-bubble--1533644108153x577480117667758100`.

URL: `https://firebase.google.com`

- **Supabase**: If you like Firebase, this is an open source alternative that provides developers with a set of tools and services for building modern web applications. It includes features such as a real-time database, authentication, file storage, and serverless functions. Supabase is built on top of PostgreSQL and offers a familiar SQL-based interface for developers. To help you integrate your app with Supabase, take a look at this plugin: `https://bubble.io/plugin/supabasejs-1671438874653x569959433685696500`.

URL: `https://supabase.io`

- **Airtable**: Airtable is a no-code tool by itself and can be used alone for various use cases; it is so versatile that the possibilities are almost endless. It combines the simplicity of a spreadsheet with the power of a database, making it easy for teams to manage projects, track workflows, and collaborate on various tasks. But you can also use it as a database for your Bubble application. You can choose to use Airtable to just store your database, or you can also use it to build a frontend for your Airtable application. Integrating it via API will allow you to show data from Airtable inside Bubble, create new records, and edit and delete existing records. It can be a powerful combination.

URL: `https://airtable.com`

- **Baserow**: Baserow is an open source online database tool that allows users to create and manage databases with ease. It offers features such as table creation, data editing, filtering, and exporting. Baserow is highly customizable and can be extended with plugins and integrations to suit various use cases. It can also be used as a database source for Bubble.

  URL: `https://baserow.io`

Some many more tools and options could be used to store your data – this list could go on and on – but the most important part is to understand that with APIs, you can integrate tools and do amazing things. If you know a tool that has an API, it can be integrated with Bubble for multiple purposes. Some other tools that are worth mentioning can be used as a database source for Bubble: SmartSuite can be an alternative to Airtable, and Notion and Coda can also be used as a database source.

You can also use well-known services used by developers such as DigitalOcean, **Amazon Web Services (AWS)**, and Microsoft Azure and use databases such as MySQL, Postgres, MongoDB, and so on.

The integration between Bubble and an external database will be made through an API, using the API Connector plugin or another specific plugin available inside Bubble to allow you to configure the integration. Once this part is configured, you will have access to the data to use on your frontend components and workflows. You can refer back to previous chapters to get more familiar with plugins, API Connector, and dynamic data inside UI components and workflows.

Basic tips to improve your app's overall performance

In the previous chapters, you learned a lot of different methods, tools, and techniques to test your app's performance. You decided to run some tests and discovered your application performance is not great. What to do now? When facing issues with performance in a Bubble app, implementing basic optimization tips can help improve the efficiency and performance of your app. Let's take a look at them:

- **Simplifying workflows**: Make sure workflows have fewer steps and run faster. Reduce the number of workflows by creating reusable ones. If an action will take too much time to load, run it in the background and let the users come back to the action later. Don't forget to properly communicate about it.

- **Minimizing database queries**: The bigger the query is, the more data you require, or the more filters you add and the bigger the data size, the more time it will take to load. Try to break queries into smaller bits of information and only query what is necessary in limited amounts. If possible, avoid big lists with too much data in them.

- **Optimizing asset loading**: Reduce the number of elements on a page. It will not only improve performance but also make it simpler for the user to interact with the interface. Make sure you optimize images. Avoid adding big images to pages and use an image optimization tool to reduce the file size; this will help with page load.

- **Reducing the number of plugins and dependencies**: Only use plugins when needed; they are a dependency and can increase the page load time as they will bring extra code into your application every time it is running. Be mindful of the plugins you use and add only essential ones; if it can be done without a plugin, try to build it natively inside your application.

- **Reducing the number of pages**: Making your application smaller doesn't mean it is less efficient or good; on the contrary. Focus on solving specific problems with a simple structure; people will thank you for that.

Additionally, you can review your app structure and ask yourself some simple questions: Can you remove a few pages? Is it possible to simplify the user flow, to make it faster and simpler to use? Can you make pages simpler and smaller and reduce unnecessary elements? If you can do it with less, you might have a winning solution; less is more. Always check if you can reduce the project scope and deliver value with fewer features and fewer pages, steps, and so on. If your app is getting too big, ask if it is actually necessary.

There will be always something to improve; this is the normal state of building software. Plan regular performance monitoring and iterative optimization and talk to your users to figure out what could be better. Not always will performance and optimization insights come from the system analytics, but they can come from users. Performing a routine for optimization is key to maintaining great app performance and ensuring a positive UX over time.

Improving app performance involves countless strategic decisions. Getting things right from the outset is more efficient than backtracking later. However, not all optimizations need to be tackled immediately; some can be addressed post-launch without compromising deadlines.

It's crucial to recognize that performance should be treated as a feature rather than the sole priority. Remember that your primary goal is to solve someone's problem, especially if you are building a start-up. Many companies start without a performing app but one that still delivers value. While prioritizing performance may be suitable for some apps, it shouldn't come at the expense of other essential steps that can move your business forward. Ultimately, getting your app to market and gathering user feedback should be the primary focus, guiding subsequent prioritization efforts.

Summary

In this chapter, we delved into optimizing performance and scalability. We explored crucial aspects of UX and design, emphasizing the importance of building user-friendly interfaces through efficient practices. You also learned about workflows, databases, and queries and how to optimize them to improve your app's overall performance. Scaling strategies and techniques were discussed to prepare your Bubble application for increased traffic. Additionally, you learned techniques to minimize page load times. Understanding these concepts is key to helping you create robust, scalable, and high-performing applications.

We have now come to the end of this book!

Here are a few tips for you to consider after finishing the book:

- Never stop learning! The journey has just begun, and there's a vast ocean of knowledge waiting to be explored. Keep pushing your boundaries and embracing new challenges.

- Put theory into action by embarking on your first real project. Applying what you've learned in a practical setting is crucial for deepening your understanding and honing your skills.

- Accept failure as an inevitable part of the learning process. Instead of dwelling on setbacks, use them as valuable learning opportunities to grow and improve.

- Celebrate successes. You deserve it. Acknowledge and celebrate your achievements, no matter how small. Recognize the progress you've made and the obstacles you've overcome and use them as fuel to propel you forward in your journey.

- Connect with like-minded individuals in the vibrant no-code community. By joining communities and engaging with other enthusiasts, you'll gain valuable insights, support, and inspiration to fuel your journey forward.

- Network strategically. Whether you are looking for friends or clients, build meaningful connections within the no-code community and beyond. Attend meetups, join online forums, and participate in networking events to expand your professional circle and learn from others' experiences.

- Experiment. There is no single path – build yours, try and test new things, break things, and keep testing new ideas and approaches.

- Set specific goals. If you are a more organized person, consider defining clear objectives for your no-code journey, whether it's building a certain number of projects, mastering a particular feature, or launching your own start-up. Having tangible goals will keep you focused and motivated. You can use the SMART goals technique.

- Dive deeper into advanced features. Now that you've grasped the basics, challenge yourself by exploring more advanced features and functionalities within Bubble. Experiment with plugins, custom workflows, and complex data structures to broaden your skill set.

- Seek feedback from friends and other no-coders. Don't hesitate to share your projects with peers, mentors, or online communities for constructive feedback. Embrace critiques as opportunities for growth and refinement and use them to iterate and improve your work.

- Document your journey. It can be just for yourself or publicly; this can help you keep track of your progress, learnings, and insights along the way. Whether it's through a blog, journal, or social media, documenting your experiences can help reinforce your understanding and serve as a valuable resource for others.

- Stay updated and informed. The world of no-code is constantly evolving, with new tools, techniques, and best practices emerging regularly. Stay informed about industry trends, attend webinars and events, and follow thought leaders to stay updated and inspired.

- Give back. Pay it forward by sharing your knowledge and experiences with others who are just starting their no-code journey. Whether it's through mentoring, teaching, or contributing to open source projects, giving back can be immensely rewarding and fulfilling. Teaching not only solidifies your own understanding but also empowers and uplifts fellow learners on their journey.

As you close the final chapter of this book, remember that you're not just a reader—you're a trailblazer in the no-code revolution. You have the power to transform your ideas into innovative solutions, without the need for extensive coding skills. So, embrace this newfound knowledge, embrace the possibilities of the no-code universe, and embark on your journey to building amazing digital applications.

I truly hope you've enjoyed this book and found it valuable in your quest to become a proficient no-code and Bubble developer. Thanks for choosing me as your guide, teacher, mentor, and friend. Now, armed with the knowledge and skills you've gained, it's time to unleash your creativity and bring your ideas to life like never before. Bring amazing projects and ideas into the world, and if you do, feel free to reach out to me and let me know! It would be very nice to know what you've been doing and have accomplished. Connect with me on my social networks (`@calderaricaio`) or visit my website (`https://www.calderari.com.br`).

Congratulations—I am proud of you!

Welcome to the no-code space; welcome to the no-code revolution!

Your journey starts now.

Further reading

- Bubble manual – performance and scaling
- `https://manual.bubble.io/help-guides/maintaining-an-application/performance-and-scaling`
- Bubble blog article – planning tips
- `https://bubble.io/blog/improve-web-app-performance/#tip-2-planning-is-essential`
- Bubble manual – *Scaling with Bubble*
- `https://manual.bubble.io/help-guides/infrastructure/hosting-and-scaling/scaling-with-bubble`
- UX book – *Don't Make Me Think* by Steve Krug
- UX book – *The Design of Everyday Things* by Donald Norman

Index

packtpub.com

Subscribe to our online digital library for full access to over 7,000 books and videos, as well as industry leading tools to help you plan your personal development and advance your career. For more information, please visit our website.

Why subscribe?

- Spend less time learning and more time coding with practical eBooks and Videos from over 4,000 industry professionals

- Improve your learning with Skill Plans built especially for you

- Get a free eBook or video every month

- Fully searchable for easy access to vital information

- Copy and paste, print, and bookmark content

Did you know that Packt offers eBook versions of every book published, with PDF and ePub files available? You can upgrade to the eBook version at packtpub.com and as a print book customer, you are entitled to a discount on the eBook copy. Get in touch with us at customercare@packtpub.com for more details.

At www.packtpub.com, you can also read a collection of free technical articles, sign up for a range of free newsletters, and receive exclusive discounts and offers on Packt books and eBooks.

Other Books You May Enjoy

If you enjoyed this book, you may be interested in these other books by Packt:

Hands-On Low-Code Application Development with Salesforce

Enrico Murru

ISBN: 978-1-80020-977-0

- Get to grips with the fundamentals of data modeling to enhance data quality
- Deliver dynamic configuration capabilities using custom settings and metadata types
- Secure your data by implementing the Salesforce security model
- Customize Salesforce applications with Lightning App Builder
- Create impressive pages for your community using Experience Builder
- Use Data Loader to import and export data without writing any code
- Embrace the Salesforce Ohana culture to share knowledge and learn from the global Salesforce community

Democratizing Application Development with AppSheet

Koichi Tsuji, Suvrutt Gurjar, Takuya Miyai

ISBN: 978-1-80324-117-3

- Discover how the AppSheet app is presented for app users
- Explore the different views you can use and how to format your data with colors and icons
- Understand AppSheet functions such as yes/no, text, math, list, date, and time and build expressions with those functions
- Explore different actions such as data change, app navigation, external communication, and CSV import/export
- Add/delete and define editing permissions and learn to broadcast notifications and inform users of changes
- Build a bot through the AppSheet Automation feature to automate various business workflows

Low-Code Application Development with Appian

Stefan Helzle

ISBN: 978-1-80020-562-8

- Use Appian Quick Apps to solve the most urgent business challenges
- Leverage Appian's low-code functionalities to enable faster digital innovation in your organization
- Model business data, Appian records, and processes
- Perform UX discovery and UI building in Appian
- Connect to other systems with Appian Integrations and Web APIs
- Work with Appian expressions, data querying, and constants

Packt is searching for authors like you

If you're interested in becoming an author for Packt, please visit authors.packtpub.com and apply today. We have worked with thousands of developers and tech professionals, just like you, to help them share their insight with the global tech community. You can make a general application, apply for a specific hot topic that we are recruiting an author for, or submit your own idea.

Share Your Thoughts

Now you've finished *Democratizing No-Code Application Development with Bubble*, we'd love to hear your thoughts! Scan the QR code below to go straight to the Amazon review page for this book and share your feedback or leave a review on the site that you purchased it from.

https://packt.link/r/1804610941

Your review is important to us and the tech community and will help us make sure we're delivering excellent quality content.

Download a free PDF copy of this book

Thanks for purchasing this book!

Do you like to read on the go but are unable to carry your print books everywhere?

Is your e-book purchase not compatible with the device of your choice?

Don't worry!, Now with every Packt book, you get a DRM-free PDF version of that book at no cost.

Read anywhere, any place, on any device. Search, copy, and paste code from your favorite technical books directly into your application.

The perks don't stop there, you can get exclusive access to discounts, newsletters, and great free content in your inbox daily

Follow these simple steps to get the benefits:

1. Scan the QR code or visit the following link:

https://packt.link/free-ebook/9781804610947

2. Submit your proof of purchase.
3. That's it! We'll send your free PDF and other benefits to your email directly.

www.ingramcontent.com/pod-product-compliance
Lightning Source LLC
Chambersburg PA
CBHW080624060326
40690CB00021B/4812